Springer Tracts in Modern Physics 93

W0111367

Springer Tracts in Modern Physics

* denotes a volume which contains a Classified Index starting from Volume 36.

B. Dorner

Coherent Inelastic Neutron Scattering in Lattice Dynamics

With 47 Figures

Springer-Verlag Berlin Heidelberg GmbH 1982

Dr. Bruno Dorner

Institut Max von Laue – Paul Langevin, B.P. 156, Avenue des Martyrs
F-38042 Grenoble, Cedex, France

Manuscripts for publication should be addressed to:

Gerhard Höhler

Institut für Theoretische Kernphysik der Universität Karlsruhe
Postfach 6380, D-7500 Karlsruhe 1, Fed. Rep. of Germany

*Proofs and all correspondence concerning papers in the process of publication
should be addressed to:*

Ernst A. Niekisch

Haubourdinstrasse 6, D-5170 Jülich 1, Fed. Rep. of Germany

ISBN 978-3-662-15776-3   ISBN 978-3-540-38572-1 (eBook)
DOI 10.1007/978-3-540-38572-1

Library of Congress Cataloging in Publication Data. Dorner, B. (Bruno). Coherent inelastic neutron scattering in lattice
dynamics. (Springer tracts in modern physics; 93). Bibliography: p. Includes index. 1. Lattice dynamics. 2. Neutrons-
Scattering. I. Title. II. Series. [QCl.S797] vol.93 [QCl76.8.L3] 539s [530.4'1] 81-14458 AACR2

© by Springer-Verlag Berlin Heidelberg 1982
Originally published by Springer-Verlag Berlin Heidelberg New York in 1982
Softcover reprint of the hardcover 1st edition 1982

2153/3130 − 5 4 3 2 1 0

Preface

The aim of this book is to present the state of the art of coherent inelastic neutron scattering as far as the investigation of lattice dynamics is concerned. Aspects of the experimental technique are discussed in much detail. Particular attention is payed to questions of resolution, intensity, focussing, and finally, optimi-
zation of the experimental setup. The treatment of the latter subject has especially
benefited from numerous discussions with scientists at the Institute Laue-Langevin,
Grenoble.

The symmetry operations contained in the space groups of the crytals under investigation play an important role in the performance of the experiment. Their influence on the analysis is discussed on experimental grounds, using examples which
avoid complicated mathematics. In several simple cases it has been possible to measure phonon dispersion curves without having to first calculate the lattice dynamical model. Yet as the number of atoms per unit cell increases, model calculations
become more and more important, and even necessary. Besides the Born-von Karman
force constant concept, particular models for ionic, metallic, and molecular crystals are presented.

The discussion of experiments starts with the information obtained from a precise determination of phonon frequencies (peak positions), and continues with a
qualitative intensity analysis of phonon peaks and an extended description of the
quantitative intensity analysis. Using the latter method, which is often called
a dynamical structure determination, the eigenvector of a particular phonon mode can
be extracted. Knowledge of eigenvectors provides a more microscopic insight into
lattice dynamics than knowing the frequencies of the dispersion curves alone does.

Several investigations of anharmonic effects follow. Generally speaking, anharmonic effects manifest themselves in the phonon lineshape and in the temperature
dependence of phonon frequencies. The usual observation is a decreasing frequency
and an increasing linewidth at higher temperatures. One particular anharmonic
effect is the soft mode observed in connection with displacive structural phase
transformations. In several cases the soft mode is accompanied by a central peak
near the phase transition. Finally, the surprising observation of a double peak for
a one-phonon response at 4.2 K is interpreted by frequency-dependent damping.

The intention of this book is to provide general information on the basis of
a detailed analysis of measurements on a restricted number of substances.

Grenoble, July 1981   *Bruno Dorner*

Contents

1. Introduction

Condensed matter appears in different states such as liquid, amorphous, and crystalline. There are substates - phases - such as superfluid liquids, the different phases of liquid crystals, amorphous states having different histories, and a very large variety of crystal structures classified into 230 space groups. There are crystalline substances which retain the same structure from lowest temperature to melting. Others undergo phase transitions from one crystalline ordered structure to another ordered one by varying, for example, the temperature. There may be partial disorder of atom positions and molecule orientations on a microscopic scale at a given temperature, such that only the averaged position or orientation is compatible with a periodic lattice. Order may appear at a lower temperature. Generally it is a question of temperature, pressure, fields, etc., and sometimes history which phase a particular material is found in.

The different phases and the transitions between them appear as a consequence of the interactions between the atoms.

There are many different techniques to study these atomic interactions. Among them, inelastic scattering of thermal neutrons has the unique advantage that thermal neutrons have wavelengths comparable to atomic distances and energies comparable to excitations in condensed matter. The technique is described in detail in Chap. 2.

The investigation of atomic interactions exhibits a many-body problem because all atoms are coupled and their displacements are not independent variables. This fact makes understanding of liquid and amorphous states extremely difficult. The analysis of inelastic neutron scattering intensities is limited to two-particle correlations as the intensity represents the squared sum over the scattered amplitudes of the different atoms. In the case of crystalline solids the many-body problem is reduced due to the periodicity of the lattice. In well-behaved crystals (away from phase transformations) translational symmetry allows restricting consideration to the smallest periodic volume, the unit cell. Additional symmetries (rotations, mirrors, etc.) facilitate the analysis of the atomic interactions further. Some basic aspects of symmetry operations and their effects in inelastic neutron scattering are discussed in Chap. 3.

In the following we will restrict ourselves to lattice dynamics in crystals, leaving out liquid and amorphous materials as well as phase transformations in solids.

These aspects have been described by SPRINGER (1972) and by LOVESEY and SPRINGER (1977). Lattice dynamics is concerned with a microscopic analysis of the different forces between the atoms. The usual procedure is to produce a lattice dynamical model with adjustable parameters which are supposed to represent the interatomic forces. These parameters are more or less plausible. Sometimes one finds that two different sets of parameters describe the experimental observation equally well. Thus the microscopic relevance of the parameters quite often remains an open question. But even a non-plausible set of parameters which describes the results of inelastic neutron scattering satisfactorily can then be used to calculate other quantities like specific heat, heat conductivity, etc. (Chap. 4).

The information one can obtain from the interpretation of the inelastic neutron scattering intensity from phonons is presented with some examples in Chap. 5. The analysis of line shapes of phonon responses yields information on anharmonic contributions as will be explained in Chap. 6.

1.1 Reciprocal Space and Normal Coordinates

As already mentioned, the atomic displacements are not independent of each other. To escape the problem of coupled coordinates one uses translational symmetry to define a reciprocal space. Points (hkl) in reciprocal space given by a reciprocal lattice vector $n \cdot \tau$ represent the set of planes in real space which is perpendicular to τ. The length of $|\tau| = 2\pi/d'$, where d' is the distance between neighbouring planes, and n is an integer. The reciprocal space is divided in many identical first Brillouin zones around each (hkl). In the following we will drop the definition "first" because lattice dynamics is only concerned with the first Brillouin zone. The second and further Brillouin zones play a role in electron band structure consideration.

To overcome the difficulty arrising from the coupling of atom displacements, one introduces normal coordinates which are plane waves in real space and represented by a wavevector q within the Brillouin zone. For one wavevector q there are 3n modes, where n is the number of atoms per unit cell. This means there are 3n dispersion branches for each direction, some of which may be degenerate. In the harmonic description these normal coordinates are orthogonal and thus uncoupled.

1.2 Momentum and Energy Transfer of the Neutron

A neutron with mass m and velocity v has a wavevector $k = 2\pi/\lambda$, where λ is the wavelength of the neutron

$$\hbar\underline{k} = m\underline{v} \ . \tag{1}$$

The direction of $\underline{k}$ is the direction of the travelling neutron, e.g., of the neutron beam. Thermal neutrons have a wavelength of about 1.8 Å, thus comparable with atomic distances. In other words, $\underline{k}$ vectors are comparable to the dimensions of Brillouin zones. The interaction of a neutron with a nucleus (LOVESEY, 1977) is described by a scattering length b and a δ-function in space at the position of the nucleus. The scattering length varies rapidly from element to element (even from isotope to isotope, most often producing unwanted incoherent scattering). In the following we consider only the coherent scattering length b_d of element d,

$$b_d = \sum_j w_j b_j \qquad (2)$$

where w_j is the probability for the scattering length b_j depending on different isotopes and different spin configurations between nuclear and neutron spin.

Incoherent scattering is considered in the following as a background which usually has a smooth Q dependence due to the Debye-Waller factor and a (sometimes disturbing) ω dependence on a spectrum related to the density of states of the sample.

The momentum transfer $\underline{Q}$ (exactly $\hbar\underline{Q}$) of a neutron in the scattering process is given by

$$\underline{Q} = \underline{k}_I - \underline{k}_F , \qquad (3)$$

where $\underline{k}_I$ and $\underline{k}_F$ are the neutron wavevectors before and after scattering. In an inelastic scattering process the energy $\hbar\omega$, transferred to the sample and lost by the neutron, is conventionally taken positive, i.e.

$$\hbar\omega = E_I - E_F = \frac{\hbar^2}{2m} (k_I^2 - k_F^2) . \qquad (4)$$

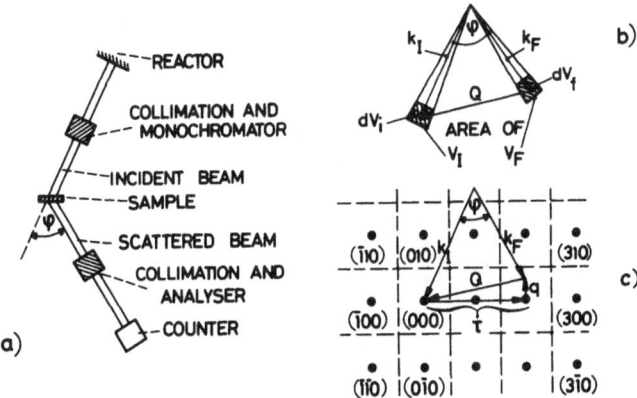

Fig. 1a-c. Inelastic neutron scattering: (a) path of neutrons in real space with "black boxes" for the determination of neutron energy before and after scattering; (b) corresponding distribution of neutrons V_I and V_F in reciprocal space around the mean wave vectors k_I and k_F; (c) momentum transfer $\underline{Q}$ of the neutron in relation to the reciprocal lattice of the sample (vectors $\underline{\tau}$) and the phonon wave vector $\underline{q}$. (DORNER and COMES, 1977)

In inelastic neutron spectroscopy we have to determine the energy E_I of the neutron before being scattered and E_F after scattering. Figure 1 is a schematic drawing of an inelastic neutron scattering experiment. The "black boxes" called collimator and monochromator and analyser determine a certain distribution of wavevector $\underline{k}_i$ around the mean or "nominal" $\underline{k}_I$, and of $\underline{k}_f$ around the mean $\underline{k}_F$. These distributions in the reciprocal space volumes V_I and V_F provide intensity (proportional to $V_I \cdot V_F$) and resolution controlled by the folding of the two distributions (DORNER, 1972).

1.3 Time-of-Flight and Three-Axis-Spectrometer Techniques

There are essentially two methods to determine the neutron energy, by

(I)   determining their velocity v in time-of-flight (TOF) technique and
(II) determining their wavelength in three-axis-spectrometer (TAS) technique.

A very detailed review on different techniques in inelastic neutron scattering has been given by DOLLING (1974). Therefore we shall keep this chapter short.

As an example for TOF we take the instrument IN 5 at the Institut Laue-Langevin in Grenoble (Fig. 2) (DOUCHIN et al., 1980). A first chopper (15,000 r/min) produces a short pulse, which spreads afterwards due to the different velocities of the neutrons. A second chopper several metres behind the first and having an electronically controlled phase with respect to the first, produces a short pulse of neutrons with a certain velocity. The resolution depends on the pulse lengths (speed of the choppers) and the distance between the choppers. The instrument IN 5 has neutron guides

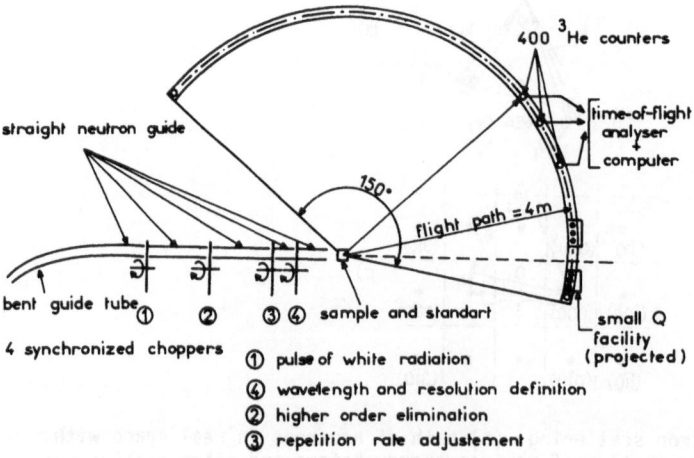

Fig. 2. Scheme of a time-of-flight (TOF) instrument (IN 5 at ILL). The choppers 2 , 3 , 4 are synchronized with respect to chopper 1. The time of flight analyser is started by a pick-up signal from chopper 4. (DOUCHIN et al., 1980)

between the choppers to avoid intensity losses with distance. A third chopper suppresses higher order neutrons having smaller velocities than the nominal ones, and a fourth chopper regulates the repetition rate of the pulses to avoid an overlap of the intensity distribution at the detectors.

Banks of detectors are installed 4 m away from the sample. The arrival time of the neutrons depends on the change in their velocity (energy) at the scattering event.

Obviously data are collected by an individual detector for a fixed scattering angle. The different energy transfers $\hbar\omega$ corresponding to the different arrival times are connected to different momentum tranfers Q.

As an example for a TAS instrument (three-axis means one monochromator axis, one sample axis, and one analyser axis, in contrast to two-axis instruments where the analyser is missing) we present the IN3, also at the Institut Laue-Langevin (Fig. 3).

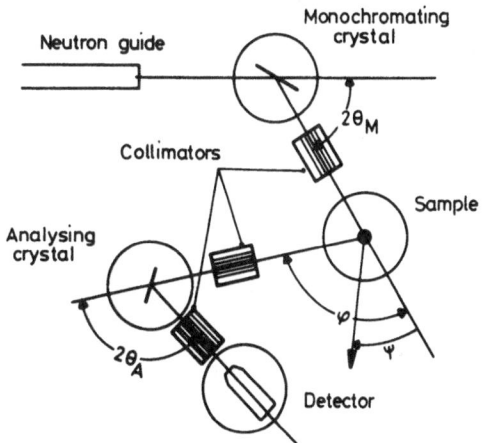

Fig. 3. Scheme of a three-axis spectrometer (TAS) (IN3 at ILL). θ_M and θ_A are the Bragg angles of monochromator and analyser, φ is the scattering angle, and ψ the orientation of the sample with respect to the incoming beam. Generally, the directions are defined by Soller slit collimators

Neutrons with a certain energy are selected by Bragg reflections from large single crystals by determination of the wavelength λ

$$2d' \sin \theta = n\lambda \quad \text{or}$$
$$2k \sin \theta = n\tau$$

(5)

where θ is the Bragg angle, d' the spacing between the reflecting planes, and $\tau = 2\pi/d'$. The higher order contaminations in the beam are accounted for by $n = 2,3,...$, (Sect. 2.2). Bragg reflection from large single crystals is used in the monochromator and the analyser. The scattering angle of the neutron beam is $2\theta_M$ and $2\theta_A$, respectively.

There are several modes of operation: constant Q scans where the energy transfer is varied, constant ω scans where Q is varied, or any combined scan. In other words, we want to determine an energy $\hbar\omega$ as a function of Q or q. In the experimental plane we have only two Q-components accessible. Therefore we have 3 unknowns, ω, Q_x, Q_y, to be determined by experiment, where we have four variables, $|\underline{k}_I|$, $|\underline{k}_F|$, the scattering angle φ, and the sample orientation ψ in the experimental plane. Apparently we can perform any scan in the ω, Q_x, Q_y space keeping one instrumental variable fixed. Most of the time $|\underline{k}_I|$ or $|\underline{k}_F|$ are kept fixed. A constant-Q scan with fixed $|\underline{k}_I|$ is shown in Fig. 4.

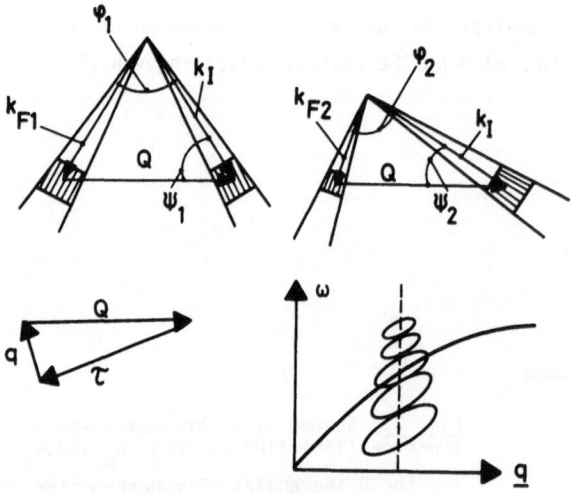

Fig. 4. Q-constant scan with k_I fixed. φ and ψ are scattering angle and sample orientation. The hetched areas give the distributions of k_i around k_I and of k_f around k_F. q is the phonon wavevector. In q-ω space a constant Q scan is drawn with varying resolution. (DORNER, 1976)

By means of peak height and width there is no preferable mode of operation. For technical reasons one sometimes chooses a scan in which the instrumental parameter variations are smallest (i.e. keep the background constant). By means of physical interpretation the constant-Q scan is a distinct one because the density of states is constant in Q space.

The TOF and the TAS techniques are complementary as TOF is best suited for incoherent scattering and in the field of coherent scattering for those problems where many data are wanted, say, knowledge of the scattering function $S(Q,\omega)$ for a large range in Q and ω. The latter is the case in liquids, for example.

Coherent scattering from single crystals is best studied by TAS technique because a variety of scans can be performed, such as constant-Q scans. For single crystals the areas of interest in Q-ω space are very much reduced by the symmetries of the crystal and the group theoretical analysis. The relevant information about lattice dynamics is concentrated on high symmetry points and lines in Q space. Working with a TAS means performing an experiment (not a measurement) since the result of each scan is extremely valuable in planning future scans. During the course of an experiment it is extremely easy to adjust energy and resolution to obtain optimum information.

These arguments are valid for continuous sources. With the construction of pulsed sources of very high peak flux TOF instruments may become favourable even for coherent scattering from single crystals.

2. Experimental Technique with Three-Axis Spectrometers

Since the first TAS was built by BROCKHOUSE (1961), the basic principle has not changed, only the mechanical and electronic engineering have been developed so that such instruments can run day and night by computer control.

2.1 Reflectivity of Monochromators and Resolution

The effectivity of a TAS depends crucially on the quality of the monochromator and analyser crystals. High quality means high-peak reflectivity and sufficiently large mosaic width. The best crystals are those for which primary extinction is negligible. If primary extinction is present (the order in the single crystal too good) then part of the volume does not participate in the reflection. Therefore, in practice monochromator crystals are chosen thicker than demanded by calculation of secondary extinction.

Very generally there is a loss of neutrons depending on the penetration depth of the neutron beam due to nuclear absorption, incoherent scattering, inelastic scattering, and parasitic reflections (DORNER, 1971).

The best material so far is pyrolytic graphite (PG) (RISTE, 1970). It has negligible primary extinction (DORNER and KOLLMAR, 1974). Nuclear absorption and incoherent scattering are negligible as well. But PG is ordered only along the c axis. It consists of stocked sheets of hexagonal graphite. These sheets are uncorrelated in the hexagonal plane. The corresponding reciprocal lattice is shown in Fig. 5. It is clearly seen that for $k \geq 1.55$ $\overset{\circ}{A}^{-1}$ parasitic reflections can not be avoided. For $k = 1.34$ $\overset{\circ}{A}^{-1}$ reflectivities of more than 90% have been observed by DORNER and KOLLMAR (1974).

With increasing k the reflectivity decreases, as more and more parasitic reflections become effective. PG also has a relative large lattice constant c: 6.71 $\overset{\circ}{A}$. Therefore the Bragg angles for higher neutron energies are quite small, thus resulting in a bad energy resolution ΔE_I,

$$\Delta E_I = 2E_I \, \Delta\theta_M \cot \theta_M \tag{6}$$

where θ_M is the Bragg angle of the monochromator, and

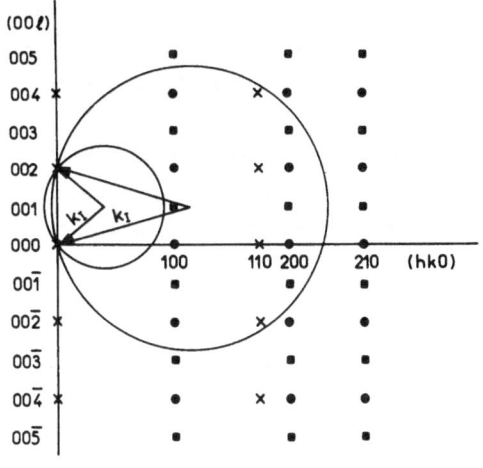

Fig. 5. Reciprocal lattice of pyrolytic graphite (PG). The positions (00ℓ) are points while all the others are rings around the c* axis. Ewald spheres for the (002) reflection are drawn for two different wavevectors k_I = 3.41 and 1.5 Å^{-1}. The Ewald spheres for wavevectors k_I > 1.5 Å^{-1} intersect with the rings of the reciprocal lattice of PG. These parasitic reflections reduce the reflectivity and cannot be avoided. (DORNER and KOLLMAR, 1974)

$$\Delta\theta_M = \left[\frac{\alpha_0^2\alpha_1^2 + \alpha_0^2\eta_M^2 + \alpha_1^2\eta_M^2}{\alpha_0^2 + \alpha_1^2 + 4\eta_M^2} \right]^{1/2} \tag{7}$$

where α_0 and α_1 are the horizontal divergencies of the beam before and after the monochromator crystal and η_M is the horizontal mosaic width of the crystal.

Inspecting (6) we see that ΔE_I depends on E_I, the Bragg angle, and on the effective horizontal divergence $\Delta\theta_M$. It is much more economic in intensity to improve the resolution for a given E_I, if desired, by using another monochromator crystal with a smaller lattice constant. In the ideal case then the intensity would decrease proportionally to the energy window ΔE_I. On the other hand, the resolution is improved by reducing $\Delta\theta_M$, then the intensity suffers quadratically because the energy window and the divergence of the beam are reduced simultaneously.

To see this effect in more detail we write the intensity of a monochromator (DORNER, 1972)

$$I_M \sim k_I^3 \cdot \frac{\alpha_0 \cdot \alpha_1 \cdot \eta_M}{\sqrt{\alpha_0^2 + \alpha_1^2 + 4\eta_M^2}} \cot \theta_M . \tag{8}$$

For simplicity we assume $\alpha_0 = \alpha_1$ and $\eta_M \gg \alpha_0$; then

$$I_M \sim k_I^3 \cdot \alpha_0^2 \cot \theta_M \tag{9}$$

and

$$\Delta E_I = E_I \cdot \sqrt{2} \, \alpha_0 \cot \theta_M . \tag{10}$$

Apparently the intensity depends quadratically on the horizontal divergence. For an optimisation of intensity (KALUS and DORNER, 1973) at a given ΔE_I it is best to make $\alpha_0 = \alpha_1$. This influence of the Bragg angle is the reason why other materials such as copper, germanium, silicon, beryllium (difficult to get good crystals) and others are used besides graphite.

Much work has been devoted in the past to the fabrication of good monochromators. For a review see FREUND (1979). FREUND (1976) developed a technique to produce controlled mosaic spread by dislocation gliding under pressure. Using this method the vertical mosaic spread can be held smaller than the horizontal one. The vertical mosaic distribution is always unwanted as it widens the beam vertically and thus causes a loss in neutrons.

2.2 Higher Order Contaminations

Besides the chosen $|k|$, the beam reflected from a monochromator contains additional higher order contamination, arising from the fact that multiples of k may be reflected by multiples of τ as long as the reactor provides a flux for the larger k's. The higher order k's are generally unwanted and are difficult to suppress. The problem can be solved only in particular cases. The simplest is to use first order $|k|$'s from the maximum of the reactor spectrum (around 40 meV). Then the higher orders come from an energy range where the reactor flux is very low. Quite often, however, one needs lower energies for better resolution.

There are crystal structures where second-order reflections have vanishing intensity, like [111] in Ge or Si.

Very often filters are used such as polycrystalline Be (VAN DINGENDEN and HAUTE-CLER, 1963) which only allows neutrons to pass with $E \leqslant 5.2$ meV, or pyrolytic graphite (MINKIEWICZ and SHIRANE, 1970) which is particularly efficient for $E = 13.7$ and 14.8 meV.

A curved (better, S-type curved) neutron guide (MAIER-LEIBNITZ, 1967; JACROT, 1970) has a natural cut-off due to curvature. For example, a neutron guide with 2700 m radius of S-type curvature does not transmit $|k|$'s larger than 3 $\overset{o}{A}{}^{-1}$ (or energies larger than 18 meV). Such a S-type curved neutron guide provides the beam from the cold source of the high-flux reactor at the Institut Laue-Langevin for the TAS-IN 12. The suppression of higher order contamination for 1.5 $\overset{o}{A}{}^{-1} < k < 3$ $\overset{o}{A}{}^{-1}$ is excellent.

2.3 Resolution and Focussing

As already shown in Fig. 1 the neutrons in the monochromatic beam have a distribution $p_i(\underline{k}_i)$ around the nominal wavevector $\underline{k}_I$. The neutrons which can pass the analyser at a given position have a distribution $p_f(\underline{k}_f)$ around the nominal analysed wavevector

$\underline{k}_F$. The transmission (or resolution) function R is the folding of the two distributions,

$$R(\underline{Q} - \underline{Q}_0, \omega - \omega_0, \underline{Q}_0, \omega_0)$$
$$= \iint p_i(\underline{k}_i) \, p_f(\underline{k}_f) \, \delta[\underline{Q} - (\underline{k}_i - \underline{k}_f)] \cdot \delta[\omega - \frac{h}{2m}(k_i^2 - k_f^2)] \, d\underline{k}_i \, d\underline{k}_f \ . \tag{11}$$

Here $\underline{Q}_0$ and ω_0 are the nominal positions of the instrument corresponding to $\underline{k}_I$ and $\underline{k}_F$. The normalisation of R is obtained by integration,

$$\iint R(\underline{Q} - \underline{Q}_0, \omega - \omega_0, \underline{Q}_0, \omega_0) \, d\underline{Q} \, d\omega = V_I \cdot V_F \ . \tag{12}$$

The measured intensity I_{meas} is

$$I_{meas}(\underline{Q}_0, \omega_0) = \iint R(\underline{Q} - \underline{Q}_0, \omega - \omega_0, \underline{Q}_0, \omega_0) \, S(\underline{Q}, \omega) \, d\underline{Q} \, d\omega \tag{13}$$

where $S(\underline{Q}, \omega)$ is the scattering function (Chap. 3).

The actual shape of R in the 4-dimensional $\underline{Q}$-ω space was first derived by COOPER and NATHANS (1967) using Gaussian approximations for the distributions. Resolution programs are now part of the standard computer libraries at reactor institutes. A graphical method should explain some basic properties of R which lead to focussing. The distribution of $p(\underline{k})$ being reflected by a single crystal monochromator exhibits a correlation between $|k|$ and the angle within the divergence controlled by a Soller slit collimator (Fig. 6a, pt.1). Somewhat simplifying (small mosaic width), the main extension is parallel to the reflecting planes. Regarding a certain path of the neutrons through the three-axis spectrometer (Fig. 6a, pt.1) in real space, we can draw Fig. 6a, pt.3 in reciprocal space showing the $\underline{k}$ distributions around $\underline{k}_I$ and $\underline{k}_F$. As the instrumental transmission or resolution function R is the folding of these two distributions, we have now, with (11), to sort out $\underline{k}_i - \underline{k}_f$ combinations for their $\Delta Q_\|$, $\Delta Q_\perp$, and $\Delta\omega$ contributions. For $|\underline{k}_i| - |\underline{k}_f| > |\underline{k}_I| - |\underline{k}_F|$, the $\Delta\omega$ is positive (energy loss of the neutron). Using only a few extreme combinations the qualitative shapes as given in Fig. 6a, pt.4 can be obtained.

We learn from Fig. 6a, pt.4 that the function R has a certain inclination in ΔQ-$\Delta\omega$ space. This inclination depends on $|k|$, on the monochromator and analyser crystals used, and on the path of the neutron, i.e. scattering to the right or to the left at M, S, or A (Fig. 6b-d).

During a constant-Q scan (Fig. 4) the transmission function varies. With k_I fixed, and varying k_F towards more energy loss, the transmission decreases as V_F decreases. Therefore phonon peaks, which have been obtained with $|\underline{k}_I|$ fixed, have to be corrected for the varying resolution during the scan. Omitting these corrections can lead to false determinations of peak centres and widths.

12

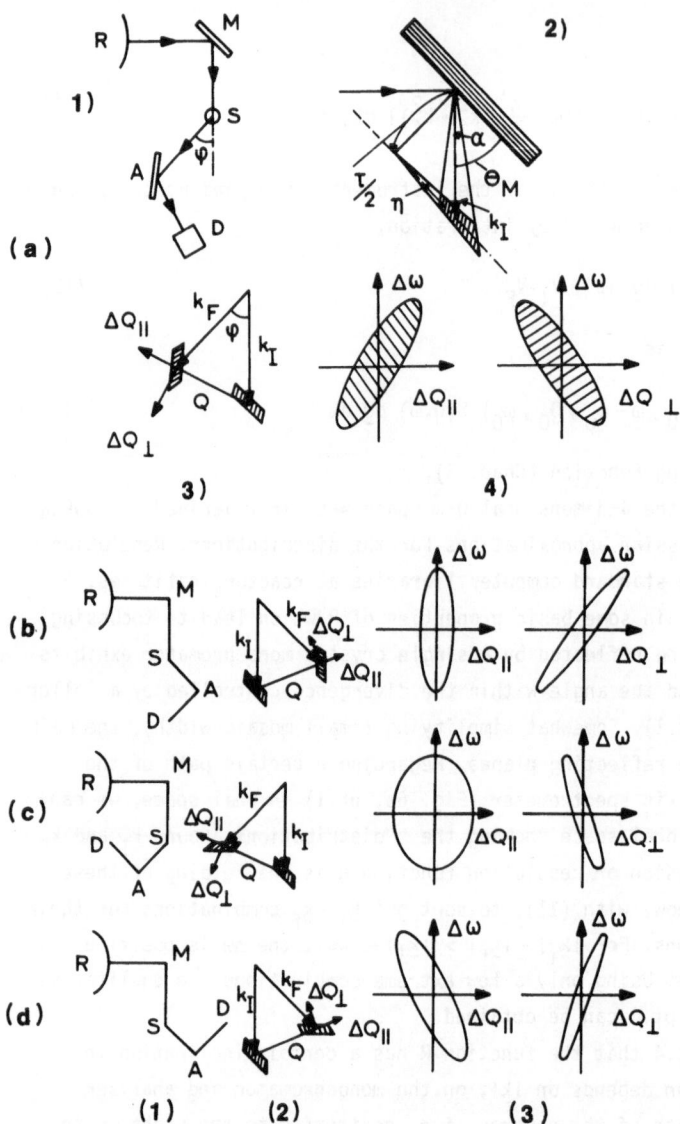

Fig. 6. (a) Resolution or transmission volume of a TAS: (1) path of the neutron beam; (R) reactor, (M) monochromator, (S) sample, (A) analyser, (D) detector; (2) reflection from a single crystal monochromator with mosaic width η and Bragg angle θ_M. The hatched area gives the distribution of $\underline{k}_i$ around $\underline{k}_I$; (3) scattering diagram for (1) in reciprocal space; (4) projections of the resolution. (DORNER, 1976). (b-d) Resolution or transmission volumes of a TAS in different geometries. Diagrams (1,3,4) are as defined in (a)

A correction, or normalization, factor $N(\omega_0) = V_I \cdot V_F$ normalizes R (DORNER, 1972). Therefore the corrected intensity I_{corr}

$$I_{corr}(Q_0,\omega_0) = \frac{I_{meas}(Q_0,\omega_0)}{N(\omega_0)} = \int \frac{R(Q-Q_0,\omega-\omega_0,Q_0,\omega_0)}{N(\omega_0)} S(Q,\omega)\, dQ\, d\omega \qquad (14)$$

represents data as if they had been obtained with a constant resolution all along the scan, yet the curve is not unfolded. This means I_{corr} still has a width, which contains the resolution. It is the integral

$$\int I_{corr}(Q_0,\omega_0)\, d\omega_0 = \int S(Q,\omega)\, d\omega \qquad (15)$$

over the pointwise corrected data which gives the integrated phonon intensity without any influence of resolution.

The $\ddot{N}(\omega_0)$ has a very simple form,

$$N(\omega_0) = C_M C_A k_I^3 \cot \theta_M \cdot k_F^3 \cot \theta_A \ . \qquad (16)$$

In the case where the resolution is much smaller than the width of the response F_j (18) (for example, a damped phonon) we can also describe the data by

$$I_{meas}(Q_0,\omega_0) = N(\omega_0)\, F_j[\omega_0,\omega_j(Q),T] \cdot |G_j(Q,Q_0)|^2 \ . \qquad (17)$$

Figure 7 shows the response of the damped (not overdamped) soft mode in $Tb_2(MoO_4)_3$ in energy gain and loss with constant k_I. The full line is a least squares fit of a $F_j(\omega,T)$ as given in (25) modified by $N(\omega_0)$. All the asymmetry is due to resolution and is well taken care of by $N(\omega_0)$. $F_j(\omega,T)$ is symmetric for such low frequencies at such high temperatures.

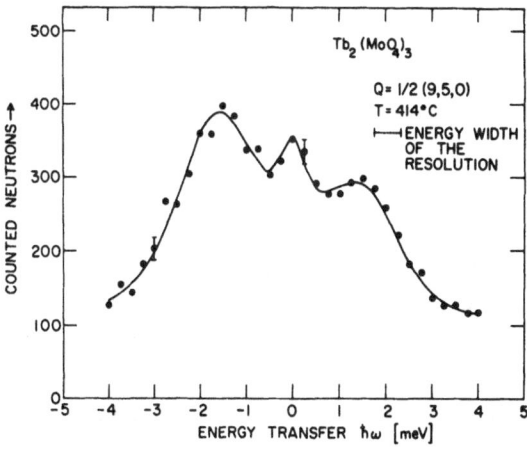

Tb$_2$ (MoO$_4$)$_3$

Q= 1/2 (9,5,0)
T= 414°C

⊢⊣ENERGY WIDTH OF THE RESOLUTION

Fig. 7. Constant Q scan through the soft mode at the M point 255°C above the phase transformation (159°C). The soft mode is heavily damped but not overdamped. The line is a least-squares fit of (25), which is symmetric in energy gain and loss. The apparent asymmetry arises exclusively from the variation of the instrumental transmission and was accounted for by including a resolution normalization factor (16) in the fit. The little maximum at zero energy transfer is background only (DORNER et al., 1972)

The resolution and normalization have so far been discussed in terms of density distributions $p(\underline{k})$ (11) and not in terms of fluxes $k \cdot p(\underline{k})$. This appears to be the appropriate choice as it corresponds to the scattering function $S(\underline{Q},\omega)$ (18) while fluxes correspond to the cross-section $d^2\sigma/d\Omega d\omega$.

When k_F is kept fixed, then a k-independent monitor in the monochromatic beam would measure the incoming flux, $k_I \cdot V_I$. But if the monitor has a $1/k_I$ characteristic, as it usually has, then it measures V_I. If data are collected, for constant monitor count rate, then $N(\omega_0)$ is taken care of by the monitor. In other words, $I_{corr} \equiv I_{meas}$. Unfortunately it is not always possible to keep k_F fixed. For example, if one uses a pyrolytic graphite (PG) filter in the incoming beam to reduce higher order contamination, then k_I has to be fixed at 2.57 or 2.67 Å^{-1} because the PG filter is most efficient for these k values.

Focussing is used to increase the peak intensity of the signal. Usually the width is reduced as well because the integrated signal intensity stays unchanged. Good focussing in Q-ω space is obtained if the ellipsoid representing the resolution function is oriented such that its long axes are as parallel as possible to the signal function, e.g. the phonon dispersion surface.

In conventional technique (flat monochromator and analyser crystals) the slope of the ellipsoid depends on the neutron energy and the scattering geometry. The variation of neutron energy is limited due to varying the intensity and, over all, resolution, and eventually by such restrictions as the use of a filter (Sect. 2.2) against higher order contamination. The influence of different geometries is of interest for the investigation of longitudinal phonons ($\Delta Q_\parallel - \Delta\omega$ plane) and for transverse phonons ($\Delta Q_\perp - \Delta\omega$ plane). A vertical orientation of the ellipsoid as in Fig. 6b,c for the $\Delta Q_\parallel - \Delta\omega$ plane would mean focussing for a longitudinal phonon dispersion curve having infinite slope, which is unrealistic. Focussing for longitudinal modes can only be obtained in the cases shown in Fig. 6a,d while focussing for transverse modes leaves the choice between all four configurations of Fig. 6. This figure visualises qualitatively different orientations of the resolution ellipsoid for different geometries and for small energy tranfers. For quantitative calculations one should use a computer program.

As long as the Bragg angle is smaller than 90^0 the long axis of the ellipsoid always has a non-zero slope in $\Delta Q - \Delta\omega$. An orientation such that the long axis is parallel to Q and the extension in $\Delta\omega$ is as small as possible would be desirable for all flat dispersion branches.

For signals centered at $\omega = 0$ but extended in Q the back-scattering spectrometer has been developed with 90^0 Bragg angles at monochromator and analyser (BIRR et al., 1971) and used with great success. But the energy resolution is too good for many physical problems, and the possible energy transfer to small.

To overcome the limitations of the back-scattering spectrometer and to obtain a horizontal orientation of the ellipsoid, sophisticated monochromators have been developed. The basic idea is to have a monochromator which reflects the same energy band everywhere within a rather large horizontal divergence. The use of horizontally curved crystals (JOHANNSON, 1933) provides such a monochromator. In this mode of operation no Soller collimators are in the beam. The resolution in energy is essentially determined by the sample size. This technique, which is very common for monochromators in X-ray scattering, was recently introduced into inelastic scattering of neutrons (MAIER-LEIBNITZ, 1972; KALUS, 1975). The resolution in momentum transfer Q is determined by the horizontal size of the monochromator crystal or by a horizontal slit close to the monochromator.

If the distances monochromator-sample and sample-analyser are fixed then the necessary curvature is a function of the neutron energy. The idea to use crystals with variable horizontal curvature as monochromator and as analyser allows obtaining any wanted slope of the resolution ellipsoid. An increase in intensity from horizontal dispersion branches up to a factor of 5 has been observed (SCHERM et al., 1977; KRAXENBERGER, 1980).

So far we have discussed focussing in reciprocal space. In real space an intensity increase can be obtained by focussing a high beam onto a small sample. This focussing is obtained by vertical curvature (CURRAT, 1973) of the monochromator. The gain in intensity comes from an increased vertical divergence, which does (in first order) not affect the energy resolution - only the vertical component of the Q resolution.

3. The Scattering Function and Symmetry Operations in the Crystal

The one-phonon scattering function for a phonon with wavevector $\underline{q}$ in branch j is given by

$$S_j(\underline{Q},\omega) = |G_j(\underline{q},\underline{Q})|^2 \, F_j[\omega,\omega_j(\underline{q}),T] \, . \tag{18}$$

The inelastic structure factor G_j

$$G_j(\underline{q},\underline{Q}) = \sum_d^{\text{unit cell}} b_d \frac{1}{\sqrt{M_d}} \, [\underline{Q} \cdot \underline{\sigma}_d^j(\underline{q})] \, \exp[-W_d(\underline{Q}) + i\underline{Q}\underline{d}] \tag{19}$$

contains the eigenvector $\underline{\sigma}$, which is normalized to unity and describes the pattern of displacements in one unit cell. It has $3n$ components, where n is the number of atoms per unit cell. The displacements $\underline{u}_d^j$ caused by the plane wave q,j are periodic through the lattice

$$\underline{u}_d^j = \frac{A^j}{\sqrt{M_d}} \cdot \underline{\sigma}_d^j(\underline{q}) \, \exp(i\underline{q}\underline{\ell}) \tag{20}$$

where A^j is an amplitude factor which depends on temperature and oscillates in time with the phonon frequency $\omega_j(\underline{q})$, $\underline{\ell}$ is the vector to the ℓth unit cell, $\underline{d}$ gives the position of atom d in the unit cell, b_d and M_d are its scattering length and its mass, and $W_d(\underline{Q})$ the exponent of the Debye-Waller factor. The inelastic structure factor G_j is the same for energy loss and gain.

The response function in energy transfer is equal to the product of the imaginary part of the dynamic phonon susceptibility χ_{Ph}

$$\chi_{\text{Ph}}[\omega_j(\underline{q}),T,\omega] = [\Omega_j^2(\underline{q}) + \Pi_j(\underline{q},T) - \omega^2]^{-1} \tag{21}$$

and the Bose occupation factor

$$\langle n \rangle = \frac{1}{\exp(\hbar\omega/kT) - 1} \qquad \text{for } \omega < 0$$

$$\tag{22}$$

$$\langle n \rangle + 1 = \frac{1}{1 - \exp(-\hbar\omega/kT)} \qquad \text{for } \omega > 0 \, .$$

It follows that

$$F_j = \frac{1}{1 - \exp(-\hbar\omega/kT)} \cdot \text{Im}\{\chi_{ph}\} \tag{23}$$

and holds for energy loss as well as for energy gain because $\text{Im}\{\chi_{ph}\}$ is an odd function in ω.

The harmonic frequency of phonon mode j is Ω_j. The simplest form of the self-energy Π_j contains a renormalization of the harmonic frequency by $\Delta_j(T)$ and a frequency proportional damping $\Gamma_j(T)$,

$$\Pi_j(\omega,\underline{q},T) = \Delta_j(\underline{q},T) - i\omega\,\Gamma_j(\underline{q},T) \tag{24}$$

Inserting (24) into (23), we find with $\omega_j^2 = \Omega_j^2 + \Delta_j$

$$F_j[\omega,\omega_j(\underline{q}),T] = \frac{\omega}{1 - \exp(-\hbar\omega/kT)} \frac{\Gamma_j(\underline{q},T)}{[\omega^2 - \omega_j^2(\underline{q},T)]^2 + \omega^2\Gamma_j^2(\underline{q},T)} \tag{25}$$

This is the response function for a damped harmonic oscillator. The first factor in (25) is called the detailed balance factor. This equation describes energy loss processes as well as energy gain processes. In the limit of vanishing damping (harmonic approximation) F_j reads

$$F_j(\omega,\omega_j(\underline{q}),T) = \frac{\omega}{1 - \exp(-\hbar\omega/kT)} \cdot \frac{\delta(\omega \pm \omega_j(\underline{q}))}{\omega^2} \,. \tag{26}$$

The $(\pm)$ sign distinguishes between energy gain and loss.

At a high temperature $(kT \gg \hbar\omega_j)$ the integrated phonon response function is

$$\int F_j[\omega,\omega_j(\underline{q}),T]\,d\omega \sim T \cdot \chi_{ph}[\omega_j(\underline{q}),T,0] = \frac{T}{\omega_j^2(\underline{q})} \tag{27}$$

proportional to temperature times the static phonon susceptibility. This means that at a high temperature the integrated intensity is the same for energy loss and gain and is independent of the damping Γ_j, even for very large Γ_j.

For $T \to 0$ or $kT \ll \hbar\omega_j$ only energy loss scattering is possible with an integrated intensity for $\Gamma_j \ll \omega_j$,

$$\int F_j[\omega,\omega_j(\underline{q}),T]\,d\omega \sim \frac{1}{\omega_j(\underline{q})} \,. \tag{28}$$

Further discussion of F_j will be given in Chap. 6. Here we shall now concentrate on the structure factor G_j.

3.1 Polarization and Symmetry of Eigenvectors

Inspecting (19) we realize that only those components of the atomic displacements are visible, which are parallel to Q.

In simple structures as copper which has fcc symmetry and contains one atom per primitive unit cell, symmetry is so high that the dispersion branches in high-symmetry directions are purely longitudinal or transvers with respect to q.

If the sample is aligned such that a [01$\bar{1}$] axis is perpendicular to the experimental plane, the corresponding plane in reciprocal space is shown in Fig. 8. The geometry for measurements of longitudinal and transverse acoustic branches is indicated. One can follow the branches from the Γ point to the X point in [100] direction. From the X point one can go in [011] direction (perpendicular to [100]) passing the K point back to Γ.

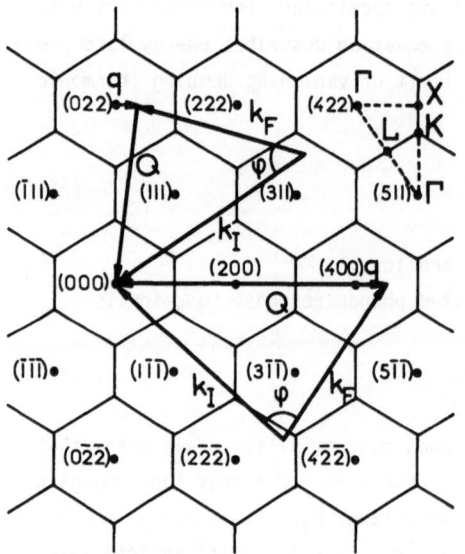

Fig. 8. Reciprocal lattice of a fcc structure with the [01$\bar{1}$] direction perpendicular to the experimental plane. Some conventional symbols for symmetry points are given in the upper right. Scattering diagrams for transverse phonons (top) and for longitudinal phonons (bottom) in [100] direction are inserted. $\underline{k}_I$ and $\underline{k}_F$ are the incoming and scattered wavevectors of the neutron, φ the scattering angle, Q the momentum transfer of the neutron, and q the phonon wavevector

Fig. 9. Phonon dispersion curves for Cu in the three major symmetry directions at 296 K. The diagram is labelled with the conventional symbols for fcc lattices. The straight lines give the initial slopes of the dispersion curves as calculated from the elastic constants. The solid lines are calculated from a Born-von Karman Model including 4th-nearest neighbours (NICKLOW et al., 1967)

Fig. 10. Phonon dispersion curves in AgBr at 85 K, (o) longitudinal, (•) transverse (1), (■) transverse(2). The broken curves are the best fit of shell model I (Table 1, p.30) (DORNER et al., 1976). Initial slopes of the acoustic branches (——) are taken from elastic constants measured at 190 K (LOJE and SCHUELE, 1970)

Fig. 11. Phonon dispersion curves of CuCl at 4.2 K. The solid lines represent the least-squares fit of a 14-parameter shell model (Table 1, p.30) (PREVOT et al., 1977)

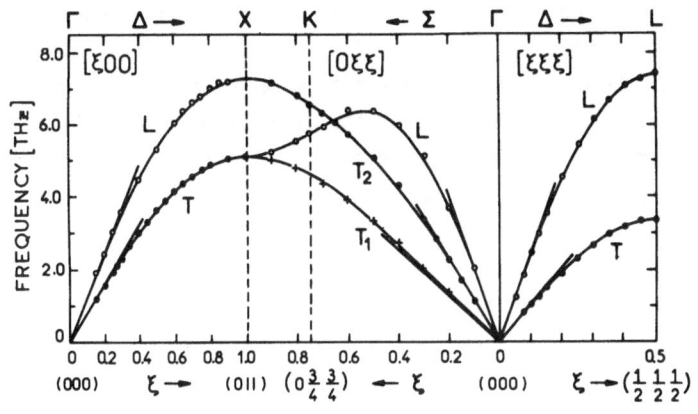

Figs.9-11.
Captions see opposite page

<u>Fig. 9</u>

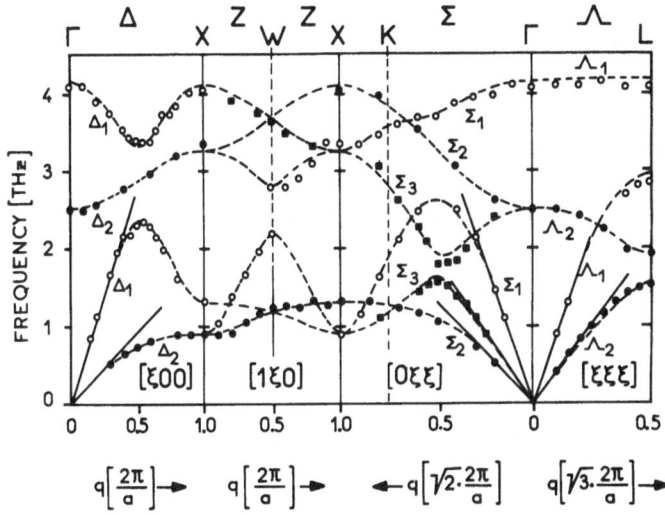

<u>Fig. 10</u>

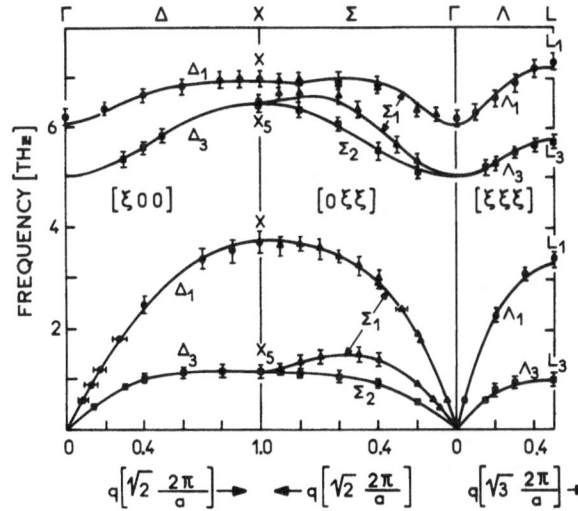

<u>Fig. 11</u>

On the way Γ-X-K-Γ the phonon dispersion curves have to be continuous. A particular X point (122) can be reached from Γ (022) in [ξ00] direction and from Γ (111) in [0ξξ] direction. At X (122) there is one non-degenerate mode polarized in [100] and two doubly degenerate modes with polarisations in the plane perpendicular to [100]. Apparently the non-degenerate mode is longitudinal as seen from Γ (022) and transverse as seen from Γ (111). The degenerate modes at X join the two degenerate transverse modes in [ξ00] direction with the longitudinal and transverse [01$\bar{1}$] in [0ξξ] direction. The dispersion curves for Cu (NICKLOW et al., 1967) are given in Fig. 9, exhibiting the connection of the dispersion branches as discussed above.

AgBr has NaCl structure (fcc) with one molecular unit per primitive unit cell, Ag at (000) and Br at (1/2 1/2 1/2). It is slightly more complicated than Cu as it has optic modes and the structure factor depends on the Brillouin zone, as will be discussed in a moment. The symmetry is still sufficiently high that the dispersion branches in high-symmetry directions are purely transverse or longitudinal. As seen from Fig. 10 (DORNER et al., 1976) the acoustic branches merge into to X point in a similar way as in Cu.

CuCl has zincblende structure (fcc) with one molecular unit per primitive unit cell, Cu at (000) and Cl at (1/4 1/4 1/4). It is again more complicated than AgBr because the symmetry is lower. In [0ξξ] direction the longitudinal mode and the transverse one (polarized in [ξ00] direction) belong to the same group theoretical representation and therefore are not pure anymore. As seen from Fig. 11 (PREVOT et al., 1977) the branch, which is predominantly longitudinal in [0ξξ] direction near Γ (highest slope) merges at X as a transverse mode to go over to the longitudinal mode for the [ξ00] direction. Along the dispersion branch there is a change from longitudinal to transverse polarization. The reason is that the two branches are in the same representation and therefore can not cross, but at the X point they are pure again as for q → 0 as well.

Note that acoustic modes which are pure in the limit q → 0 where elasticity theory is applicable, are not necessarily pure for finite q's. Generally speaking for crystals with several atoms per unit cell, if there are atoms at general positions in the unit cell, then the branches are not anymore purely transverse or longitudinal in the sense that $\underline{\sigma}$ contains displacements only perpendicular or parallel to $\underline{q}$. Nevertheless there may be several group theoretical representations; none of them is pure.

As far as acoustic branches are concerned it is often convenient to call them "longitudinal" and "transverse", although they are not pure. In α-quartz (9 atoms per unit cell), for example, there exists a 2-fold axis in [110] direction and consequently two representations, one symmetric and one antisymmetric. The "longitudinal" acoustic branch in this direction belongs to the symmetric representation, and the two "transverse" ones to the antisymmetric representation. In Fig. 12 we give typical eigenvectors for the three Si atoms. There is no general extinction for the symmetric

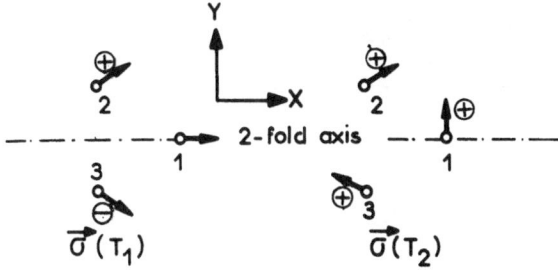

Fig. 12. The two types of eigenvectors σ belonging to the two representations T_1 and T_2 in the presence of a twofold axis. The possible atomic amplitudes in the x-y plane are given by arrows and the z components by + and − signs. The T-K-M direction and the M point in α-quartz are a typical example. In the quartz structure Si atom (1) lies on the twofold axis, while Si atoms (2) and (3) are symmetry related. For the representations see Table 2, p.62, where $m_1 \longleftrightarrow T_1$ and $m_2 \longleftrightarrow T_2$

representation. But all modes belonging to the antisymmetric representation are extinct if $\underline{Q}$ is parallel to $\underline{q}$ (ξξ0). This is easily seen from (19) because I) the amplitude of Si(1) is perpendicular to $\underline{Q}$ and is thus invisible, and II) the phase factors $\exp(i\underline{Q}\underline{d})$ for Si(2) and Si(3) are the same but the components of their amplitudes parallel to $\underline{Q}$ are opposite, thus their contributions cancel. This fact was used to identify three branches of the symmetric representation in α-quartz (DORNER et al., 1980a).

If the symmetry operation is not a simple twofold axis as in quartz but a twofold screw axis as in naphthalene along the b axis, then there exist extinction rules for both representations for $\underline{Q}$ parallel to the [0ξ0] direction. The symmetric modes are extinct in (0k0) zones for odd k's and the antisymmetric for even k's. The reason is that two symmetry-related atoms are displaced by b/2 in b direction. Therefore the phase between the two differs by π·k, thus the phasefactor $\exp(i\underline{Q}\underline{d})$ is opposite for the two symmetry-related atoms for odd k and equal for even k. The extinction appears everywhere from the zone centre to the zone boundary, where one symmetric and one antisymmetric mode are degenerate (Fig. 13) (NATKANIEC et al., 1980). This particular phenomenon of extinction related to screw axis or glide planes will be discussed in more detail in Sect. 3.3.

3.2 Intensity of Phonons in Different Brillouin Zones

As mentioned above, Brillouin zones are different with respect to phonon intensity. Even Bragg intensities are different due to interference, proportional to $(b_{Ag} + b_{Br})^2$ for even Miller indices and $(b_{Ag} - b_{Br})^2$ for odd indices in AgBr. In CuCl there are three types of Bragg intensities: $(b_{Cu} + b_{Cl})^2$ for $h + k + \ell = 4n$, $(b_{Cu} - b_{Cl})^2$ for $h + k + \ell = 4n + 2$, and $(b_{Cu}^2 + b_{Cl}^2)$ for odd indices.

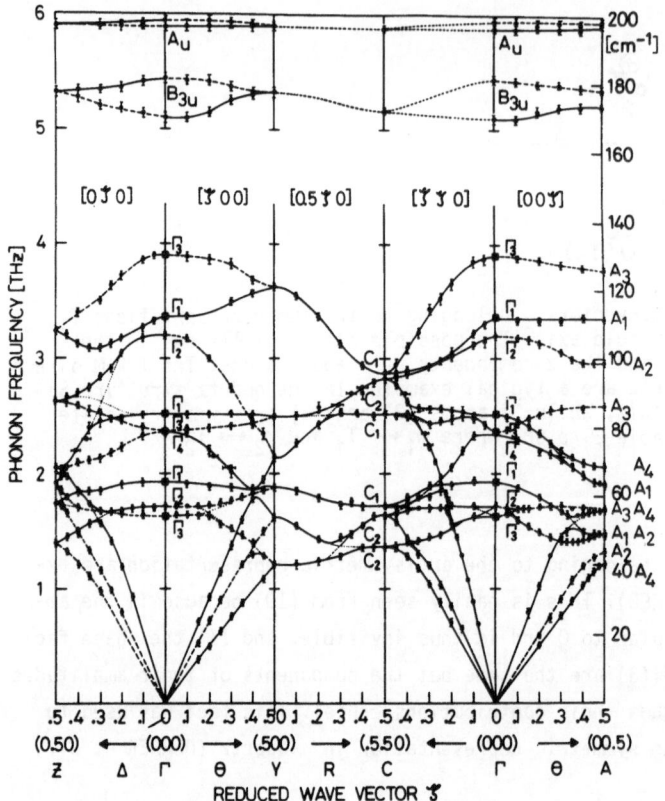

Fig. 13. Experimental phonon dispersion curves in naphthalene at 6 K. Different representations are shown by full and dashed lines and at the zone centre and zone boundaries by labels Γ_i, C_i and A_i. The full and dashed lines denote symmetric and antisymmetric phonon branches with respect to the twofold screw axis in [0ξ0] direction, and to the glide plane in [ξ00] and [00ξ] directions. The crossings of eigenvectors (anticrossing of branches) are indicated by dotted lines. Most of them have been verified experimentally. All lines are guide lines to the eye. For calculated curves see Fig. 20. (NATKANIEC et al., 1980)

To calculate phonon intensities near the Brillouin zone centre the same expressions enter as above, only the scattering length b_d is "decorated" by $\underline{q} \cdot \underline{\sigma}_d^j(\underline{q})$. In systems with two atoms per unit cell and for $|q| \to 0$ the eigenvector of an acoustic branch becomes $\underline{\sigma}_1 \approx \underline{\sigma}_2$ (exactly: $\underline{u}_1 = \underline{u}_2$ and $\sigma_1 = \sqrt{M_1/(M_1 + M_2)}$, $\sigma_2 = \sqrt{M_2/(M_1 + M_2)}$), and of an optic branch $\underline{\sigma}_1 \approx \underline{\sigma}_2$ (exactly: $M_1 \underline{u}_1 = - M_2 \underline{u}_2$ and $\underline{\sigma}_1 = \sqrt{M_2/(M_1 + M_2)}$, $\underline{\sigma}_2 = - \sqrt{M_1/(M_1 + M_2)}$). Thus Brillouin zones of type $(b_1 \underline{q} \underline{\sigma}_1 + b_2 \underline{q} \underline{\sigma}_2)^2$ (strong Bragg intensities) are favourable for the investigation of acoustic modes, and type $(b_1 \underline{q} \underline{\sigma}_1 - b_2 \underline{q} \underline{\sigma}_2)^2$ (weak Bragg intensities) for optic modes. These simple rules are valid only near the zone centre. As an example we present in Fig. 14 the intensity variation in AgCl predicted from model calculation by VIJAYARAGHAVAN et al. (1970). AgCl has fcc symmetry with two atoms per unit cell as AgBr.

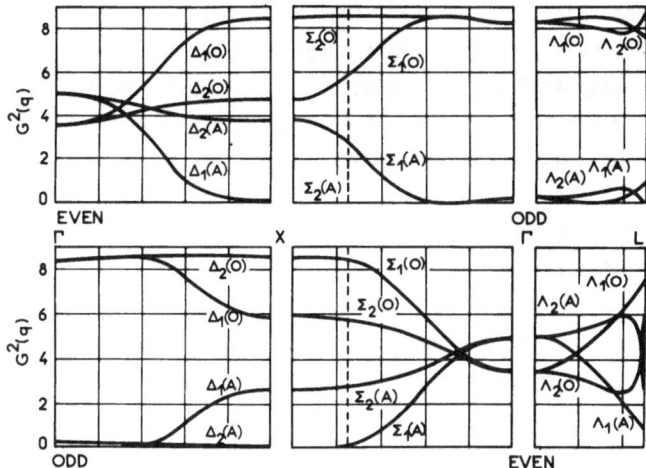

<u>Fig. 14.</u> Calculated inelastic structure factors squared G^2(q) for AgCl on the
basis of a shell model (19). EVEN and ODD stands for Brillouin zone centres with
all even or all odd indices (VIJAYARAGHAVAN et al., 1970). The labels for the
different branches can be compared to Fig. 10 (AgBr). The G^2(q) in AgBr have been
found very similar to those in AgCl (DORNER et al., 1976)

The structure factor G_j and thus predictions of phonon intensities get much more
complicated as the number of atoms per unit cell increases. It would be a hopeless
waste of time to try to investigate a complete set of phonon dispersion curves in a
complex system without precalculated intensities from a lattice dynamical model.

3.3 Extended Zone Scheme for Non-Symmorphic Space Groups

In symmorphic space groups where the space group symmetries can be explained by point
group symmetries, the dispersion curves have to have a horizontal tangent $\partial\omega/\partial q$ at
the zone boundary. This is not necessarily the case in non-symmorphic space groups
which contain screwaxes and glideplanes. For directions in which the unit cell is
enlarged by the translation included in the screwaxis or glideplane the phonon
branches at the zone boundary are degenerate. Pairs of branches which become degenerate merge into the zone boundary with finite slope, one having the opposite slope
of the other (WINSTON and HALFORD, 1949).

The simplest non-symmorphic space group is D_{6h}^4 (Pb$_3$/mmc) in which we find the
hcp monatomic substances such as many metals. The positions of the two atoms are
(000) and (1/3 2/3 1/2) or (1/3 2/3 1/4) and (2/3 1/3 3/4) in hexagonal coordinates.
Due to the presence of the second atom, the unit cell is enlarged only in the c direction. Resuming some arguments already pointed out in Sect. 3.2 we find that the

Bragg intensity at (001) is proportional to $(b-b)^2$; this means extinct. From the same arguments we find that the optic mode is visible proportional to $(b+b)^2$ at (001). Although the Bragg intensity at (001) is extinct, the Brillouin zone boundary is at (0 0 1/2). Lattice dynamical calculations for the longitudinal modes in c direction with $Q \parallel q$ show that in Brillouin zones (00ℓ) with even ℓ, only the LA branch is visible while the LO branch is extinct, and vice versa for odd ℓ. Apparently the lattice dynamical calculation for the [00ξ] direction reflects only one atom per unit cell. In other words, the offset of the second atom in the hexagonal plane does not come into play for $q = [00\xi]$. In the case of a monatomic substance in hcp structure, the extended zone scheme - plotting the acoustic and the optic mode side by side and not on top of each other - is very appropriate as the dispersion curves become smooth curves without any irregularity at the zone boundary (but a finite slope), and most important as the observable intensity follows the extended zone scheme. In practice one would measure the longitudinal dispersion curve from (002) to (003) obtaining first the LA and then the LO part. All along the structure factor G_L is proportional to $(b+b)^2 = 4b^2$. One could also say it is as if there would be only one atom per unit cell with $G_L \sim b^2$. In this case the square of the sum over all unit cells would be 4 times bigger.

There are glideplanes and screwaxes, in the hexagonal plane in D_6^4 as well but the periodicity is not increased by these symmetry operations. Therefore, the extended zone scheme has no meaning for directions in the hexagonal plane, and the dispersion curves have horizontal tangents at the zone boundaries.

A more complicated system is naphthalene, which has a monoclinic structure with the non-symmorphic space group P 2_1/a. It contains 2 molecules per unit cell. One molecule can be transformed into the other by a screwaxis along b or by a glideplane in the a-c plane with translation in the a direction. In both directions [ξ00] and [0ξ0] we find the degeneracy at the zone boundary as discussed above, but not for the [00ξ] direction because both possible translations are perpendicular to [00ξ]. Neglecting the rotation in the screw and the mirror in the glideplane, we could assume that the unit cell has half the volume containing only one molecule. The axes a and b would be halved as periodicity distance, but would not be coordinates of the unit cell anymore. The new coordinates a'b' would be a kind of diagonal between a and b assuring that $a \cdot b = 2 \cdot a' \cdot b'$.

During the measurements of the dispersion curves of naphthalene (BOKHENKOV et al., 1977; NATKANIEC et al., 1980) the extended zone scheme was extreme useful in particular for the determination of the librational branches which will be discussed in more detail in Sect. 4.3.

4. Lattice Dynamical Models

The oldest idea to produce a lattice dynamical model is the BORN-Von KARMAN (1912) method using central forces between the atoms. There is always a lowest number of forces necessary to stabilize the structure. To describe phonon dispersion curves usually the number of forces (parameters) had to be increased including interactions between more distant atoms. This increase of the number of parameters was often felt to be unsatisfying because a better fit to the experimental data did not provide a more profound understanding of the basic physics.

The Born-von Karman model has been described extensively by BORN and HUANG (1956). The following assumptions and approximations form the basis of this theory and the others which will be presented in the following sections.

a) The adiabatic approximation. The electrons are always able to adapt themselves to the instantaneous nuclear positions. Thus the potential energy may be written as a general Taylor series in terms of the displacements of the atoms from their equilibrium positions.

b) The harmonic approximation. The atomic displacements u are considered to be so small that the above series expansion may be broken off after the quadratic term.

c) The requirement of periodic boundary conditions. This is equivalent to replacing the finite specimen by an infinite medium without boundary effects.

The zeroth term in the Taylor expansion gives the lattice energy in equilibrium. The first term has to vanish to garanty equilibrium. Let us call the second term (BILZ and KRESS, 1979)

$$\phi_2 = \frac{1}{2} \sum \underline{u}(L) \; \underline{\phi}(L,L') \; \underline{u}(L') \tag{29}$$

where $L = (\ell,d)$ denotes the dth particle in the ℓth cell. The 3×3 two ion force constant matrices $\phi(L,L')$ are subject to the conservation laws of energy, momentum, and angular momentum and to the symmetry restrictions of the space group.

From the vector equation of motion

$$M_d \; \ddot{\underline{u}}(L) = - \frac{\partial \phi_2}{\partial \underline{u}(L)} = - \sum_{L'} \underline{\phi}(L,L') \; \underline{u}(L') \tag{30}$$

one obtains the eigenvalues and eigenvectors of the system as the solutions $\omega_\lambda^2 = \omega_j^2(\underline{q})$

and $\underline{\sigma}_\lambda = \underline{\sigma}^j(\underline{q})$ of the secular equation

$$\det |\underline{D}(\underline{q}) - \underline{M}\omega^2\underline{I}| = 0 \; , \tag{31}$$

where the dynamical matrix $\underline{D}$ is defined as

$$\underline{D}(dd'|\underline{q}) = \sum_\ell \underline{\phi}(L,L') \; \exp\{-i\underline{q}[\underline{R}(L) - \underline{R}(L')]\} \tag{32}$$

and $\underline{M} = (M_d)$ is a diagonal mass matrix, while $\underline{I}$ is a unit matrix $\underline{I} = \delta_{\alpha\beta} \cdot \delta_{dd'}$, where $\alpha\beta$ are Cartesian indices. The index j labels the different branches belonging to every wavevector $\underline{q}$.

These equations represent lattice dynamics theory in harmonic approximation. As long as the parameters (the two ion force constants in a Born-von Karman model) are not calculated ab initio by microscopic theory, they are determined by fitting the model to experimental observations such as phonon dispersion curves. Experimental observations at a given temperature already contain anharmonic effects such as thermal expansion. Therefore the set of fitted parameters will be different for different temperatures. Allowing for a temperature dependence of the parameters is called the quasi-harmonic approximation.

A model which reproduces the measured frequencies satisfactorily does not necessarily elucidate the real microscopic interactions. Szigeti (LEIGH et al., 1971) pointed out that very different sets of parameters for a Born-von Karman model describe the dispersion curves equally well. But the eigenvectors which are calculated by the different sets are different. A comparison to measured eigenvectors (Sect. 5.3) or to second-order Raman scattering provides a further criterion on the physical relevance of a chosen set of parameters.

Besides these weak points in the Born-von Karman model, this model was even unable to reproduce particularities of phonon dispersion branches like the LO-TO splitting in ionic crystals or the dispersion in metals where the inclusion of force constants between up to 8th nearest neighbours increased the number of parameters (> 30) but not the quality of the fit (BROCKHOUSE et al., 1961).

During the last 20 years a tremendous effort has been made to improve the theory of lattice dynamics [see (HORTON and MARADUDIN, 1974) and references therein]. Models have become more sophisticated and more microscopic, but they still provide only a parametrisation of observable quantities. For developements including anharmonicity and lattice defects see LUDWIG (1967).

4.1 Ionic Crystals: AgBr and CuCl

Long-ranged Coulomb forces play a major role in ionic crystals with the ionic charge $Z \cdot e$. The displacements of the ions produce electric dipoles and eventually macroscopic electric fields which are responsible for the LO-TO splitting. The polarisability of the ions is included in the shell model (DICK and OVERHAUSER 1958; COCHRAN 1959c) where part of the outer electrons with the charge Y·e are given an amplitude $\underline{w}$ for motion against the core. The potential depends then in a bilinear form on all $\underline{w}(L)$ and $\underline{u}(L)$ of the lattice. The equations of motions are

$$M_d \, \ddot{\underline{u}}(L) = - \frac{\partial \phi_2}{\partial \underline{u}(L)} \tag{33}$$

as (30) for the ions and

$$M_{e\ell} \, \ddot{\underline{w}}(L) = - \frac{\partial \phi_2}{\partial \underline{u}(L')} = 0 \tag{34}$$

for the electrons. This equation represents the condition of the adiabatic approximation for the shell model, i.e. the mass of the electron is set to zero.

One has to introduce Coulomb and short-range forces between ions, between electrons, and between ions and electrons. The dynamical matrix is then written as

$$\underline{D}(\underline{q}) = \underline{Z}\underline{C}(\underline{q})\underline{Z} + \underline{R}(\underline{q})$$
$$- [\underline{T}(\underline{q}) + \underline{Z}\underline{C}(\underline{q})\underline{Y}] \, [\underline{\bar{S}}(\underline{q}) + \underline{Y}\underline{C}(\underline{q})\underline{Y}]^{-1} \, [\underline{T}^{+}(\underline{q}) + \underline{Y}\underline{C}(\underline{q})\underline{Y}] \; . \tag{35}$$

The ion and the shell charges $\underline{Z}$ and $\underline{Y}$ are written in matrix form. $\underline{C}(\underline{\bar{q}})$ is the matrix describing the Coulomb interaction taking into account the different coupling to the macroscopic electric field for $q \to 0$ depending on the polarization relative to $\underline{q}$ as derived by KELLERMANN (1940). The matrix of short-range forces (repulsive between nearest neighbours) is $\underline{R}(\underline{q})$; $\underline{T}$ and $\underline{\bar{S}}$ represent the short-range shell-core (different ions) and shell-shell coupling. The matrix $\underline{\bar{S}}$ consists of the short-range electron-electron coupling $\underline{S}$ and the shell-core coupling $\underline{K}$ (inside one ion) (Fig. 15).

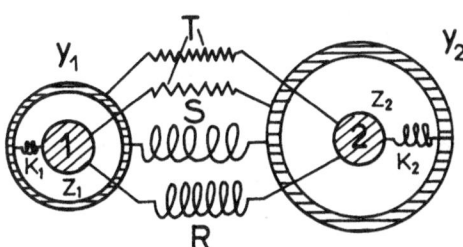

Fig. 15. The shell model for ionic crystals.

This shell model contains the following parameters to describe ionic crystals with two atoms per unit cell:

Z: the ionic charge +Z on the cations and -Z on the anions
(the label 1 corresponds to the cation and 2 to the anion);

Y_1, Y_2: shell charges;

K_1, K_2: force constants between shell and core;

(In the following the letter A represents a longitudinal force constant and B a transverse force constant).

A^R_{12}, B^R_{12}: force constants between nearest cores;

A^T_{12}, B^T_{12}: force constants between core and shell of neighbouring atoms (usually one takes $A^T_{12} = A^T_{21}$; $B^T_{12} = B^T_{21}$);

A^S_{12}, B^S_{12}: force constants between nearest shells;

A_{11}, B_{11}: force constants between next nearest neighbours (cations);

A_{22}, B_{22}: force constants between next nearest neighbours (anions).

Among the six nearest-neighbour force constants one is redundant (WOODS et al., 1960). Quite often one uses $A^R_{12} = A^T_{12}$ (DOLLING and WAUGH, 1965; PRICE et al., 1971). Thus the shell model with shells on each ion contains 14 parameters. As will be discussed in the following some constants may have values near to one another. Then with some physical justification one will set them equal to reduce the number of free parameters. Some of the parameters may turn out to have negligible influence on the quality of the fit; thus one keeps them constant, sometimes zero.

The parameters for the shell charge Y_i and for the shell-core coupling are very often replaced by the electrical α_i and the mechanical d_i polarizabilities,

$$\alpha_i = \frac{Y^2_i}{K_i + \alpha} \quad ; \quad d_i = -\frac{\alpha Y_i}{K_i + \alpha} \tag{36}$$

where α is a combination of short-range force constants and depends on the structure. For AgBr, $\alpha = A^R_{12} + 2B^R_{12}$, and for CuCl, $\alpha = 4A^T_{12} = 4A^R_{12}$.

Many crystals have both an ionic and a covalent contribution to their bindings [see, for example, the ionicity scale by (PHILLIPS, 1970)]. Only the alkali halides can be considered purely ionic crystals.

AgBr crystallizes in the NaCl structure, which is typical for ionic interactions, but in the description of the lattice dynamics some covalent effects have been discovered which are described by quadrupolar force constants. The local virtual excitation of a 4d electron on the silver ion into an excited s state leads to a quadrupolar compressibility on the Ag ion. A nearest neighbour d-p hybridization leads to a quadrupolar rotatory term.

The phonon dispersion curves have been measured by DORNER et al. (1976), (Fig. 10). The drawn curves are the result of shell model calculations including quadrupolar terms as introduced by FISCHER et al. (1972). The parameters of the model were fitted to the experimental data by minimizing the χ^2 function:

$$\chi^2 = \frac{1}{N - K} \sum_i \left(\frac{\nu^i_{obs} - \nu^i_{calc}}{\sigma_i} \right)^2 \tag{37}$$

where N and K are the number of phonons and parameters, respectively, and σ_i is the experimental error of the ith phonon. The final values of the parameters are listed in Table 1, together with the standard error

$$SE = \left[\frac{1}{N - K} \sum_i \left(\nu^i_{obs} - \nu^i_{calc} \right)^2 \right]^{1/2} . \tag{38}$$

The low value obtained for χ^2 indicates that the experimental error bars are about two times larger than the deviations of the calculated dispersion curves from the experimental points. The experimental error bars given include the error from counting statistics, obtained by a least-squares fit to the phonon group, and a systematic uncertainty allowing for slight misalignments of sample and instrument. The latter can only be estimated from experience and is usually accounted for by taking twice the standard deviation. For the fit of the model, therefore, the most important factor regarding the experimental errors is that they scale relatively for different phonons. In table 1 we present the parameter sets of three different models for AgBr and of one model for CuCl. In the case of AgBr we have a good example to illustrate that this model fitting procedure is first hand a parametrisation of the experimental data to permit further calculations such as the density of states, specific heat, etc. In model I the Ag ion is polarisable; in model II it is not. The change in χ^2 is a factor of 2, which would be called a significant improvement if $\chi^2 > 1$. But in this case χ^2 is considerably smaller than 1. This means that the experimental error estimates (as mentioned above) are larger than the deviations of the calculated curve from the measured points. Suppressing the polarisability of the Ag ion the 9 parameters left free adapt new values – some of them considerably different as the polarisability of the Br ion and the quadrupole terms $S \Gamma^+_{12}$ on the Ag ion.

These two quadrupolar terms contain the essential difference between the models applied by DORNER et al. (1976) and the model applied by FUJII et al. (1977). The experimental data in both cases are comparable. Although χ^2 is bigger in the work of Fujii et al., this is partly due to the fact that these authors quote smaller experimental errors than Dorner et al. A value for SE is not given. Let us assume that the difference in the quality of the fit is not significant. But there are significant differences. Fujii et al. include 15 free parameters, 6 more than Dorner et al. in model II. Some of them are physically unreasonable. They find a repulsive force K_1 between core and shell of the Ag ion, and B^T_{12} and B^S_{21} should be negative quantities similar to B^R_{12}.

Table 1. Parameters in shell model calculations. Sets of parameters are given as obtained by fitting such models to experimental data

	Units	AgBr DORNER et al. (1976) I	AgBr DORNER et al. (1976) II	AgBr FUJII et al. (1977)	CuCl PREVOT et al. (1977)
Z	e	0.880	0.928	1.09	0.353
α_1	A^3	1.996	-	-0.062	0.0025
d_1	e	-0.04	-	-0.0740	0.0440
α_2	A^3	3.027	3.497	6.40	0.074
d_2	e	0.106	0.149	0.130	0.1321
A^R_{12}	$e^2/2v$	11.697	11.615	12.02	11.255
B^R_{12}		-1.737	-1.906	-3.50	11.587
B^T_{12}		B^R_{12}	B^R_{12}	-10.675	10.380
B^T_{21}		B^R_{12}	B^R_{12}	1.372	B^T_{12}
A^S_{12}		A^R_{12}	A^R_{12}	21.40	23.057
B^S_{12}		B^R_{12}	B^R_{12}	2.21	24.205
A_{11}		-	-	-0.431	0.3210
B_{11}		-	-	0.081	0.0350
A_{22}		2.364	2.330	2.52	0.2628
B_{22}		-0.195	-0.272	0.187	0.3192
$S\Gamma^+_{12}$		1.054	0.581	-	-
$S\Gamma^+_{15}$		0.229	0.288	-	-
Numbers of Parameters		11	9	15	14
SE	THz	0.06	0.09		0.07
χ^2		0.14	0.32	1.38	
Y_1		7.94		0.09	-0.066
$\dot{Y}_2$		-5.36	-3.88	-5.25	-0.625
K_1		1543.0		-11.13	22.32
K_2		406.0	196.0	202.7	163.6

Comparing the models we conclude that the 9 (or 11) parameter model proposed by FISCHER et al. (1972), who introduced two quadrupolar terms to account for the covalent part of the 4d electron configuration on the Ag ion, is physically more meaningful than the pure shell model with 15 parameters. The Fischer et al. model gives more microscopic insight into the interactions between the ions in AgBr, but nevertheless derived quantities such as density of states, specific heat, etc., are equally well calculated from all fitted models.

The experimentally determined phonon dispersion curves in CuCl (zink-blende structure) were fitted by a 14 parameters shell model (Fig. 11) (PREVOT et al., 1977). Apparently the short-range order forces play a dominant role, while the Coulomb interactions are less strong than in AgBr, as can be seen from the smaller ionic charge and the lower polarisabilities. Already from the structure we can conclude that CuCl has partly covalent character of binding typically for the tetrahedral configuration of anions around a cation and vice versa. If the Coulomb interactions are dominant we find the NaCl structure with octahedral coordination. In Fig. 11 there are no experimental results quoted for the TO mode near the Γ point. In this region particular anharmonic effects have been observed (HENNION et al., 1979) which will be discussed in detail in Sect. 6.3.

4.2 Metals: Cadmium

Soon after the first measurements of dispersion curves in metals (BROCKHOUSE et al., 1961) it was realized that Born-von Karman models are unappropriate since they are unphysical and are not able to describe the measured dispersion curves satisfactorily even including many neighbouring atoms in the calculation. It is now generally recognized that it is more satisfactory to formulate models in which conduction electron-phonon interactions are explicitly dealt with. The Hamiltonian of a metal can be written as (SHAM and ZIMAN, 1963)

$$H = H_e(\underline{r}) + H_i(\underline{R}) + H_{e,i}(\underline{r},\underline{R}) \tag{39}$$

where $H_e(\underline{r})$ is the Hamiltonian of the electrons with electron coordinates $\underline{r}$, including electron-electron interaction, $H_i(\underline{R})$ is the Hamiltonian of the ions with position vectors $\underline{R}$ and $H_{e,i}(\underline{r},\underline{R})$ is the electron-ion coupling term.

Separating the degrees of freedom for ions and electrons we describe the electrons in the field of fixed ion positions $\underline{R}$. We write the Schrödinger equation for the electrons,

$$[H_e(\underline{r}) + H_{e,i}(\underline{r},\underline{R})] \; \psi_m(\underline{r},\underline{R}) \;\; = \;\; E_m(\underline{R}) \; \psi_m(\underline{r},\underline{R}) \tag{40}$$

where $E_m(\underline{R})$ are the energy eigenvalues in the electronic states $\psi_m(\underline{r},\underline{R})$ at fixed ion positions $\underline{R}$.

A total wave function Ξ_n for the whole crystal satisfying the equation

$$H \, \Xi_n = E_n \, \Xi_n$$

is then

$$\Xi_n(\underline{r},\underline{R}) = \sum_m \phi_{n,m}(\underline{R}) \, \psi_m(\underline{r},\underline{R}) \ .$$

The wave functions $\phi_{n,m}$ must satisfy the equation

$$[H_i(\underline{R}) + E_m(\underline{R})] \, \phi_{n,m}(\underline{R}) + \sum_{m'} (A_{mm'} + B_{mm'}) \, \phi_{nm'}(\underline{R}) = \mathscr{E}_n \, \phi_{n,m}(\underline{R}) \ . \tag{41}$$

The adiabatic approximation (Chap. 4) consists in ignoring contributions from non-diagonal components of $A_{mm'}$, $B_{mm'}$. The diagonal components themselves are negligible compared with H_i and E_m. For a large part of the calculation of phonon dispersion curves it is sufficient to include only the ground-state energy $E(\underline{R})$ of the electron system (at fixed $\underline{R}$). The influence of the electron distribution in the states near the Fermi level is usually small and in the following encluded in the susceptibility (45).

The Schrödinger equation for the whole crystal is then simplified. After Fourier transformation we write

$$[T_i(\underline{q}) + U_i(\underline{q}) + E(\underline{q})] \, \phi_n(\underline{q}) = \mathscr{E}_n \, \phi_n(\underline{q}) \ . \tag{42}$$

Here $T_i(\underline{q})$ is the kinetic energy of the ions, and $U_i(\underline{q})$ the ion-ion potential energy of the positive ions immersed in a uniform compensating background of negative charge. $E(\underline{q})$ are the energy levels of the electrons, and $\mathscr{E}_n$ is the energy of the crystal. To calculate the energy $E(\underline{q})$, one starts by describing one conduction electron in the field of a positive ion with his core electrons. This is done by introducing model potentials. We use here the local model potential (ABARENKOV and HEINE, 1965) (Fig. 16a).

Outside of the core we have normal Coulomb attraction Ze^2/R, where Z gives the ionisation of the ion. The parameter R_M resembles the radius of the core electron distribution, and A is a parameter for the depth of the potential inside the core. The potential is taken isotropic. There has been theoretical effort to calculate appropriate values R_M and A from the exact wave function in the depletion hole of the core (SHAW and HARRISON, 1967). We take the two parameters as model parameters to be determined by fitting experimental data.

Local model potentials work rather well for non-transition metals because their cores are compact. For transition metals the cores are much more extended and core-core interactions have to be expected. The model potential in Fig. 16a can be written in Fourier transform

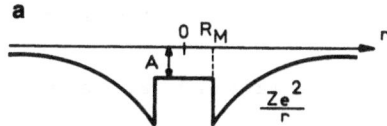

a

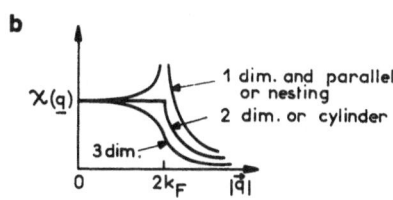

b

Fig. 16a,b. Some details of pseudopotential models for metals. (a) The local model potential after ABARENKOV and HEINE (1965) experienced by a free electron in the presence of a positive metal ion. R_M and A are adjustable parameters to take the core electrons into account. (b) The susceptibility $\chi(q)$ of free electrons (45) for different dimensions of the electron system (Di SALVO, 1977). k_F is the Fermi wavevector

$$V(q) \sim -\frac{1}{q^2}\left[(1+A)\cos(qR_M) - A\frac{\sin(qR_M)}{qR_M}\right]\exp[-\xi(\frac{q}{2k_F})^4] . \tag{43}$$

The parameter ξ which cuts off the osciallations at large q is only of small influence.

The potentials of the other ions are included by the structure factor $S(q)$

$$S(\underline{q}) = \frac{1}{N}\sum_{\underline{R}}^{N} e^{-i\underline{q}\underline{R}} . \tag{44}$$

The other electrons screen the potential which is described by a dielectric screening function $\varepsilon(\underline{q})$. The electrons interact with each other as included by the electron susceptibility $\chi(\underline{q})$. Without electron exchange and correlation the single electron susceptibility is

$$\chi(\underline{q}) = \frac{f(E_{\underline{k}}) - f(E_{\underline{k}+\underline{q}})}{E_{\underline{k}+\underline{q}} - E_{\underline{k}}} \tag{45}$$

where E is the electron energy for wavevectors $\underline{k}$ and $\underline{k}+\underline{q}$ and f is the density of states corresponding to these energies. In Fig. 16b $\chi(\underline{q})$ is shown for electronic systems of different dimensions; $\underline{k}_F$ is the electron wavevector on the Fermi surface.

In perturbation calculation we find (HARRISON, 1966)

$$E(\underline{q}) \sim |S(\underline{q})|^2 \frac{V^2(\underline{q})}{\varepsilon(\underline{q})} \chi(\underline{q}) . \tag{46}$$

The total energy $\mathscr{E}$ of the crystal is the sum of the ionic and the electronic energy.

$$\mathscr{E}_n = U_i + E . \tag{47}$$

This energy depends on the positions of the ions $\underline{R}$. At T = 0 the ions are supposed to occupy their ordered equilibrium positions $\underline{R}_0$ (disregarding the zero point motion). This state, in which no phonons are excited, has the lowerst energy of the crystal $\mathcal{E}_0$.

If we introduce, by a *gedankenexperiment*, one static plane wave with wavevector $\underline{q}_0$ and amplitude $\underline{u}(\underline{q}_0)$

$$\underline{u}(\underline{q}_0) = \underline{u}_0(\underline{q}_0) \cdot \sin(\underline{q}_0\underline{R}) ,$$

the resulting distortion of the lattice will increase the crystal energy to $\mathcal{E}_n$. This change in energy $\mathcal{E}_n - \mathcal{E}_0$ is the phonon energy $\hbar\omega(\underline{q}_0)$ in case that the distortion is not static but dynamic and if the amplitude $\underline{u}_0(\underline{q}_0)$ corresponds to one quantum $\hbar\omega(\underline{q}_0)$ (one phonon). The distortion increases the ion energy U_i and decreases the electron energy E such that the phonon energy may be the small difference between large quantities.

It is in the framework of the adiabatic approximation that the distortion is introduced as static. The distortion enters the electron energy (46) by the structure factor

$$S(g) = \frac{1}{N} \sum_{R}^{N} e^{i\underline{g}\underline{R}} = \frac{1}{N} \sum_{R}^{N} \exp\{i\underline{g} [\underline{R}_0 + \underline{u}(\underline{q}_0) \sin(\underline{q}_0\underline{R})]\} , \qquad (48)$$

which provides a strong contribution at $\underline{g} = \underline{q}_0$.

Anomalies in the phonon dispersion curves - Kohn anomalies (KOHN, 1959) - are expected where the phonon wavevector $\underline{q}$ is equal to $2k_F$ modulo a reciprocal lattice vector $\underline{\tau}$,

$$|\underline{q} + \underline{\tau}(hk\ell)| = 2k_F . \qquad (49)$$

These anomalies are caused by a drastic change of $\chi(\underline{q})$ in (45) and Fig. 16b. In the 3-dimensional case $\chi(\underline{g})$ has a vertical tangent at $2k_F$. Near a Kohn anomaly we can write the phonon frequency $\omega(\underline{q})$

$$\omega^2(\underline{q}) = \Omega^2(\underline{q}) [1 - \lambda(\underline{q}) \chi(\underline{q})] \qquad (50)$$

where $\Omega(\underline{q})$ is the phonon frequency without electron-phonon coupling, and $\lambda(\underline{q})$ the electron-phonon coupling parameter. Thus the phonon coupling reduces the phonon frequency for $|\underline{q} + \underline{\tau}| \lesssim 2k_F$ (for details see Sect. 5.1).

The dispersion curves of the non-transition metal Cd have been measured by DORNER et al. (1981c). Since naturally occurring cadmium strongly absorbs neutrons (2650 barns), this experiment has been performed with a monocrystal of the isotope ^{110}Cd.

Among the non-transition metals, zinc and cadmium are of particular interest owing to the high anisotropy of their crystal structures: their hexagonal lattice is greatly stretched along the c axis (c/a = 1.86 and 1.89, respectively, compared with the ideal value of 1.623). This makes most of the physical properties of these metals much more complicated and particularly interesting to study.

From the point of view of the modern theory of metals both elements are of interest because they are intermediate between simple (s - p) and transition metals.
Although the d shells of Zn and Cd atoms are completely filled, the energy of the
d electrons is close to the boundary of the conduction band and, therefore, they
can appreciably influence the metallic binding energy. The effect of d electrons is
to produce a complicated electron-ion interaction in Zn and Cd. Therefore the at-
tempts to describe their properties quantitatively within the framework of the simple
pseudopotential model, that works well, for example, in the case of alkali metals
(VAKS et al., 1978 and references there), are not very successful.

The obtained phonon spectrum of cadmium is shown in Fig. 17. The dashed lines
are drawn through the experimental points by eye, while the solid lines are the calculated dispersion curves using the local model potential of Abarenkov - Heine type
with a few fitting parameters which are explained later. The spectrum obtained is

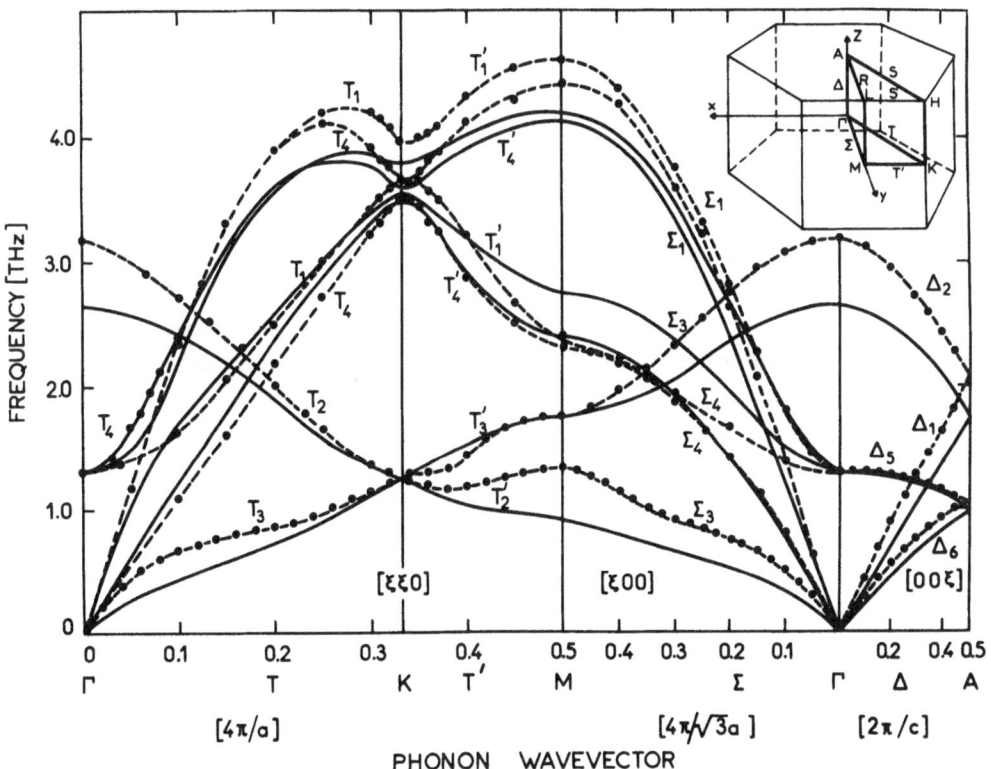

Fig. 17. Phonon dispersion curves in Cd at 80 K. Dashed lines are drawn by eye
through the experimental points. The solid lines represent calculations with a
local Abarenkov - Heine model potential. The Brillouin zone and symmetry labels for
a hcp structure are given in the insert (DORNER et al., 1981c)

stable for all wave vectors. The accuracy of the frequency reproduction at the boundary of the Brillouin zone is, on the average, about 10%. However, to describe high-
frequency and low-frequency branches simultaneously with reasonable accuracy is difficult. The calculations are not adequate because the potential taken in accordance
with the phonon spectrum does not result in the correct axial ratio c/a, but corresponds to a structure close to the ideally packed one.

The calculations performed show that the allowance for the higher orders of perturbation theory within the framework of the potential model chosen does not appreciably improve the description of the structure and the phonon spectrum as a whole.
It is not clear as yet what can be gained in this respect by introducing the non-
local pseudopotential that, according to BERTONI et al. (1974), substantially affects
the structural characteristics of the cadmium lattice. The present paper considers
how the model based on the Abarenkov – Heine potential (43) can be applied to describe
the finer features of the cadmium phonon spectrum that are discussed below.

Instead of parameter A it is more convenient to use the value of wavevector q_0
corresponding to the first zero of the potential which is related to the well depth
by the expression

$$A = \frac{q_0 R_M \cdot \cos(q_0 R_M)}{\sin(q_0 R_M) - q_0 R_M \cdot \cos(q_0 R_M)} . \tag{51}$$

For the calculations presented in Fig. 17 $q_0 = 1.6\ k_F$ and $R_M = 0.8$ Å were used.
Correlation and exchange interaction between the electrons were taken into account
using the tabulated values of TOIGO and WOODRUFF (1970). The parameters have been
chosen such that for the lowest frequency at the K point both experimental and calculated frequencies coincide. The cut-off parameter ξ did not vary much and was
0.02.

It follows from the modern theory of non-transition metals that the non-pair interionic interaction, mediated by the conduction electrons, must influence the formation of the equilibrium structure and phonon spectrum of the metal. One would expect this influence to be particularly pronounced in anisotropic lattices such as
Zn and Cd. However, as has already been pointed out, this influence has not been
revealed so far by analysing the general information on the spectrum and structure.
But in hexagonal metals there is a specific possibility of directly revealing the
non-pair forces that was considered in detail by BROVMAN et al. (1969).

Pair interactions between ions result in axial symmetry of the force matrix for
every ion pair. As a result, the dynamical matrix can possess a higher symmetry
than is determined by the space group of the lattice. This can lead to additional
symmetry in the phonon spectrum at certain points of the Brillouin zone. However,
the symmetry becomes lower when non-pair forces are present. The behaviour of the
phonon frequencies in hexagonal metals at symmetry point K of the reciprocal lattice
can be an example of such a situation.

A group theoretical analysis shows that the phonon frequencies of the two T_1 and two T_4 branches at point K are determined in terms of the dynamical matrix elements by the expressions:

$$\omega_1^2 = \text{Re}\{D_{11}^{xx}\} - \text{Im}\{D_{11}^{xy}\}$$

$$\omega_2^2 = \text{Re}\{D_{11}^{xx}\} + \text{Im}\{D_{11}^{xy}\} + 2\,\text{Re}\{D_{12}^{xx}\}$$

$$\omega_3^2 = \text{Re}\{D_{11}^{xx}\} - \text{Im}\{D_{11}^{xy}\} \tag{52}$$

$$\omega_4^2 = \text{Re}\{D_{11}^{xx}\} + \text{Im}\{D_{11}^{xy}\} - 2\,\text{Re}\{D_{12}^{xx}\} .$$

If non-pair interaction is absent, then $\text{Im}\{D_{11}^{xy}\} = 0$; in this case the squares of the frequencies ω_2^2 and ω_4^2 are at the same distance from the square of the degenerate frequency $\omega_1^2 = \omega_3^2$. When the non-pair forces exist, the values $\omega_1^2 = \omega_3^2$ is shifted with respect to $(\omega_2^2 + \omega_4^2)/2$ by the value $2 \cdot \text{Im}\{D_{11}^{xy}\}$. Different approaches of the dispersion curves T_1 and T_4 to point K can be realized depending on the relation between the elements of the dynamical matrix (Fig. 18a). To estimate the effect of non-pair forces on the Cd phonon spectrum, the behaviour of the dispersion curves near point K was thoroughly investigated. Note that phonon measurements at point K are very difficult because of the rapid change of the polarization vectors and, therefore, of the neutron inelastic scattering cross-section that takes place close to K. This may result in somewhat larger experimental errors in determining the frequencies near point K than in other regions of the Brillouin zone. Nevertheless,

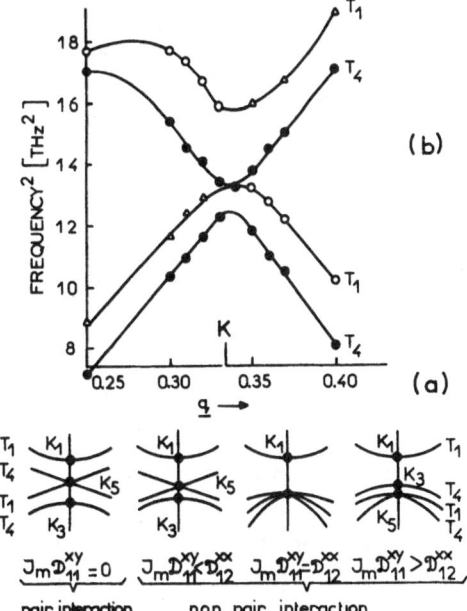

Fig. 18a,b. The splitting of frequencies in Cd at the K point. (a) Different theoretical predictions as a function of additional non-pair interactions; (b) experimental values of the squares of phonon frequencies near the K point (DORNER et al., 1981c)

the data obtained (Fig. 18) indicate that the case $Im\{D_{11}^{xy}\} < Re\{D_{12}^{xx}\}$ is realized in cadmium, that is, the non-pair forces are of importance in determining the phonon spectrum, and they should be accounted for in a correct description of the properties of this metal.

The measurements revealed a curious feature of the cadmium phonon spectrum: the existence of a contact point of the T_1 branches belonging to the same irreducible representation. Such accidental degeneracy (i.e. not due to lattice symmetry) is un-usual in the phonon spectra of single crystals. In all the experimental and calcu-lated phonon spectra of hcp metals reported so far, the branches T_1 are separated by a noticeable gap. Therefore, we considered it interesting to investigate the de-generate region in more detail, with better resolution and accuracy. The results are shown in Fig. 19 where along with the dispersion curves the variation of the

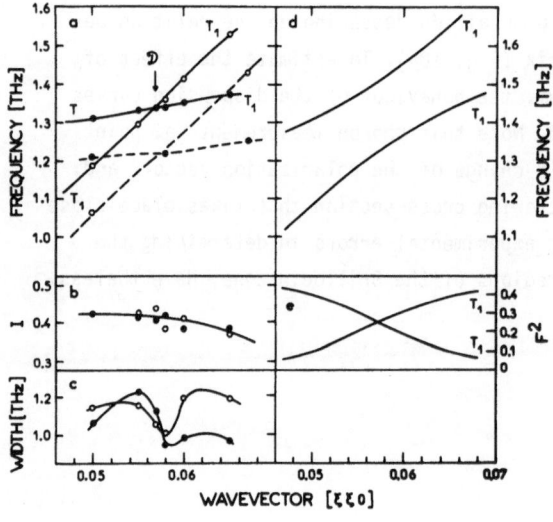

Fig. 19a-e. The accidental dege-neracy of two T_1 branches at point D. (a) $\nu(q)$ measured at 80 K (———) and at 300 K (----); (o) measurements in the direction $[-(1+\xi),(2+2\xi),0]$, (●) measure-ments in the direction $[(2+\xi),-2\xi,0]$; (b) integrated intensi-ties of phonon groups at 80 K; (c) change of phonon width near point D; (d) $\nu(q)$ calculated by the Abarenkov-Heine model poten-tial; (e) behaviour of the cal-culated inelastic structure fac-tor squared G^2 for the two branches in (d) for the direction $[(2+\xi),-2\xi,0]$ (DORNER et al., 1981c)

integrated intensities and widths of the phonons near the point of degeneracy are shown. Analysis of the results shows that when $q = q_d = 0.057$ the frequencies of the two branches coincide to within 0.01 THz.

The behaviour of the scattered intensity determined by the dynamic structure factor $F(q)$ is also evidence for the degeneracy. Fig. 19e shows how $F(q)$ should be-have in one of the directions studied when the degeneracy is absent; both branches should be observed near q_d with gradually varying intensity. When degeneracy of the frequencies takes place, a rapid renormalization of the polarization vectors will occur at q_d, and only one branch with about a constant intensity will be observed in experiment. It is this case that is realized in practice (Fig. 19b). The behav-iour of the widths near q_d is not of monotonic character (Fig. 19c), but the oscillations are not large and are probably due to changing focusing conditions of the spectrometer. Changing the temperature to 300 K does not result in any appreciable anomalies in the phonon widths at point q_d and only changes the general level of the frequencies (Fig. 19a). These observations indicate that no pecularities of the phonon-phonon interaction arise at the degeneracy point.

To investigate possible reasons for the degeneracy we performed calculations of the phonon spectrum near point q_d by varying the parameters of our model potential q_0 and R_M, changing the axial ratio c/a and taking into account both the pair and three-particle interionic interactions via the conducting electrons. The position of point q_d changed a little, but in all cases the gap between the branches remained considerable, no less than 0.15 THz. Fig. 19d shows the calculation which gives the best fit over the entire phonon spectrum. Thus, to explain the observed feature of the cadmium spectrum a more refined model is required.

As a result of the experiments performed, cadmium can be considered one of the metals whose phonon spectrum has been studied most thoroughly. However, this is not the case for the theoretical description of its properties. We have examined the possibilities of the conventional pseudopotential model using a simple form of the Abarenkov - Heine local potential. The only complication of the theory was the allowance for the non-pair interaction via conduction electrons.

When describing the phonon spectrum as a whole, this approach provides reasonable results for many problems, reproducing the phonon frequencies in the symmetry directions with an accuracy no worse than 10%. As far as a correct description of the essential features of the spectrum is concerned, the model proves to be inadequate. The need for comprehensive improvement of the calculation scheme becomes evident. In addition to the non-pair interactions, the non-local form of the potential connected with the role of d electrons should be taken into account, and a proper allowance should be made for the feature of its band structure connected with a considerable deviation of the real Fermi surface from sphericity.

It must be expected that the strong anisotropy of the lattice and other specific properties of cadmium cannot be described at all using the spherically symmetric potential of the electron-ion interaction and that the anisotropy of this interaction in such metals will have to be considered. Our studies of the strength of the Kohn anomalies (Sect. 5.1) indicate this need.

4.3 Molecular Crystals: Naphthalene and Anthracene

Molecular crystals are built from units which contain many atoms generally covalently bound. Usually these units, the molecules, are very stable in solid, liquid, and gaseous phases. The forces between them are weak. They are typically of van der Waals type with I) an attractive part from induced dipol moments proportional to r^{-6} and

II) a repulsive part proportional to r^{-12} (Lennard-Jones potential) or proportional to $\exp(-\alpha r)$ (Buckingham potential). These potentials have been used very successfully in calculations of the lattice dynamics of rare-gas crystals.

KITAIGORODSKII (1961) developed the idea to describe the forces between the molecules by a sum of Buckingham potentials between atoms of different molecules

$$\phi = \sum_{i,j} \phi_{ij} = \sum_{i,j} - \frac{A_{ij}}{r_{ij}^6} + B_{ij} \exp(-\alpha_{ij} r_{ij}) \tag{53}$$

where i and j are labels of different atoms. For each pairing of atomic species he took a special parameter set A, B, α. This ansatz turned out to be very successful for organic molecular crystals, which contain mainly hydrogen and carbon. There are 9 adjustable parameters for the three pairings H-H, H-C, C-C.

KITAIGORODSKII (1966) and WILLIAMS (1966, 1967) derived parameter sets using information from selected sets of crystal structures, heats of sublimation, and other thermodynamic data. The obtained parameter sets for particular atomic pairings have been found transferable to a large variety of molecular substances. They could be used to predict structures of molecular crystals.

To find the stable crystal structure, minima in the total potential energy of the crystal are sought allowing 6 degrees of freedom per molecule, 3 translations and three rotations. This assumption of rigid molecules is a valid approximation only if the frequencies of a free molecule (internal modes) are higher than the frequencies of translations and librations in the solid (external modes). The general condition for equilibrium (KITAIGORODSKII, 1966) is that the first derivatives of the total potential energy $\phi(u_i)$, in (53), with respect to each structural parameter u_i including the lattice parameters should be zero,

$$\frac{\partial \phi}{\partial u_i} = 0 . \tag{54}$$

Usually it is sufficient to include first- and second-neighbour molecules into the calculation.

PAWLEY (1967, 1972) developed a computer program to calculate phonon dispersion curves of molecular crystals. He used both parameter sets KP (Kitaigorodskii) and WP (Williams). The frequencies calculated with the two sets came out similar. After very recent inelastic neutron scattering experiments we have the impression that WP describes the measured dispersion curves better than KP.

At the time the calculated frequencies could only be compared with results from light scattering, mainly Raman data. The agreement (about 20% discrepancy) was astonishingly good considering that the calculations were based on parameters obtained from macroscopic observations averaged from several materials. For coherent inelastic neutron scattering from molecular substances one needs fully deuterated single crystals of low mosaic spread and sufficient volume. The measurements demand good resolution in ω as well as in q because there are many branches near each other and the size of the Brillouin zone is small. This simply comes from the fact that molecules are big units and the unit cell contains several of them, thus the unit cell is large and the sum of the translational and librational degrees of freedom of all molecules in the unit cell gives the large number of dispersion curves.

Molecular crystals are soft materials which show strong phonon damping at room temperature. Not all crystals can be cooled to a low temperature for investigation because many undergo phase transformations connected with so drastic changes in volume that the single crystals crack when going through the phase transformation.

Extended and very precise measurements of phonon dispersion curves have recently been performed on naphthalene (BOKHENKOV et al., 1977; MACKENZIE et al., 1977; NATKANIEZ et al., 1980) at 6 K. The sample was very slowly cooled, 5 K/h down to 200 K and 10 K/h lower. The experimental results for naphthalene at 5 K are presented in Fig. 13. For details about the experimental procedure see Sect. 5.2. The calculated dispersion curves are shown in Fig. 20. The qualitative agreement in shape between the calculated and the measured curves is surprisingly good, but quantitatively some frequencies are off by about 20%. In particular, the separation of Γ_3 (3.90 THz) and Γ_1 (3.37 THz) is not completely reproduced and the splitting be-

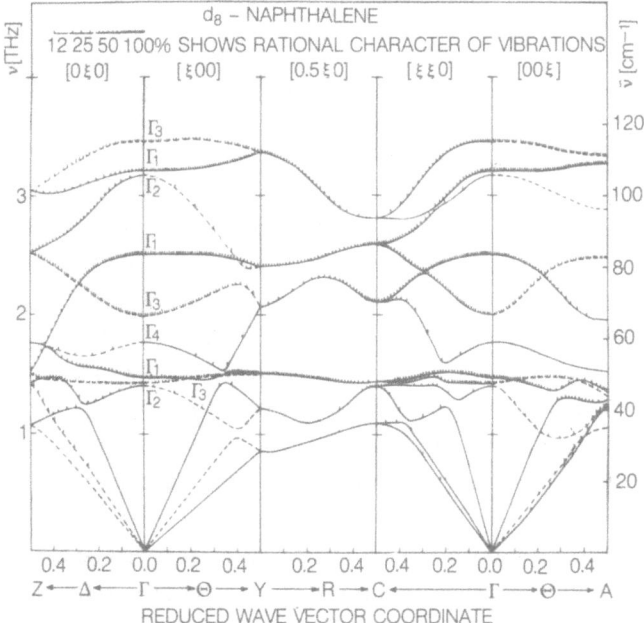

Fig. 20. Calculated phonon dispersion curves for naphthalene with the Buckingham potential (53) using WP. For experimental curves see Fig. 13. The full and dashed lines denote symmetric and antisymmetric phonon branches with respect to the two-fold screw axis in [0ξ0] direction and to the glide plane in [ξ00] and [00ξ] directions (NATKANIEC et al., 1980)

ween Γ_1 (2.52 THz) and Γ_3 (2.38 THz) comes out too large. This defect of the model can be improved by introduction of a multipole moment (RIGHINI et al., 1980).

Dispersion curves in naphthalene have been measured at 100 K under pressure P of 3 Kbar and compared to the data at 100 K at normal pressure (SCHMELZER et al., 1980). Under pressure the frequencies had increased by 10% on the average. Experimental Grüneisen parameters have been derived and compared with calculated ones.

In the course of applying pressure on a molecular crystal the six unit cell constants a , b , c , α , β , γ change, as do the molecular positions and orientations. The only changes which are associated with a change in unit cell volume are those which involve the unit cell parameters. Repositioning or reorientation of the molecules does not cause a change in volume and therefore there are no non-zero linear terms in the energy expansion involving these coordinates. The non-zero linear terms are of the form

$$\partial\phi = P \, \partial V \tag{55}$$

where P is the hydrostatic pressure. ∂V is the change in volume due to the change ∂u_i in u_i, where u_i is any one of the unit cell parameters:

a , b , c , α , β , γ -   triclinic case

a , b , c , β -   monoclinic case (naphthalene and anthracene)

a , b , c -   orthorhombic case .

The calculation was performed in the following way (PAWLEY and MIKA, 1974)

$$\frac{\partial\phi}{\partial u_i} - P \frac{\partial V}{\partial u_i} = 0 . \tag{56}$$

Solving (56) one finds a new equilibrium structure which has smaller unit cell constants. The new positions and orientations of the molecules inside the new unit cell are found by minimizing the lattice energy as described before. In general, the distances between the molecules are reduced. Calculation of the dispersion curves in this equilibrium structure under pressure gives higher frequencies. The calculated Grüneisen parameters for naphthalene agree very well with the measured ones.

A volume change appears under pressure (decrease) *and* with temperature (increase). Some dispersion curves in naphthalene have been determined at different temperatures (SHEKA et al., 1981). As (56) can be used for negative hydrostatic pressure as well, we calculated phonon frequencies for increasing volume (temperature does not enter the calculation). The calculated phonon frequencies decreased more rapidly with increasing volume than the measured one with increasing volume (produced by temperature) (Fig. 21). From this observation we conclude that frequencies of naphthalene determined at constant volume would increase with temperature. In this context we like to mention that ZALLEN and CONWELL (1979) have claimed that at constant volume the frequencies of organic molecular crystals increase and those of inorganic molecular crystals decrease with increasing temperature.

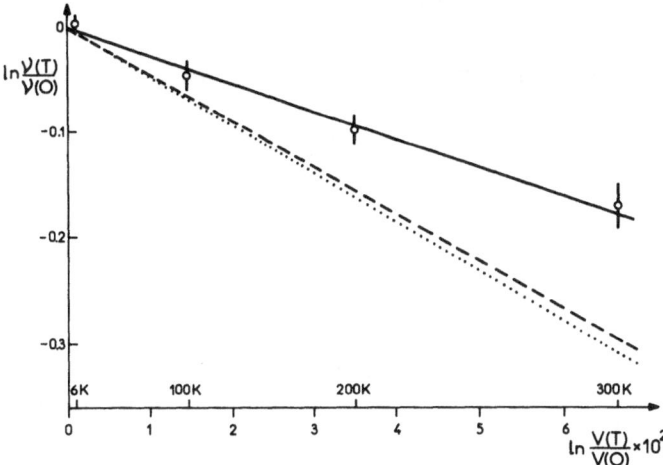

Fig. 21. Frequency change of the 2.4 THz libration in naphthalene with increasing volume. The experimental points ϕ are obtained at normal pressure with increasing temperature (indicated on the abscissa). The dashed and dotted curves are calculated at constant temperature and negative hydrostatic pressure using the measured and the calculated mode Grüneisen parameters, respectively. (SCHMELZER et al., 1981)

Recently we performed inelastic neutron scattering on anthracene (a flat molecule of three benzene rings) (DORNER et al., 1981b) having the same space group as naphthalene with two molecules per unit cell. This case is more complicated than naphthalene because the lowest internal modes lie in the range of the external mode frequencies. To include the internal modes into the calculations one follows the idea of perturbation theory (TADDEI et al., 1973). One solves first the free molecule dynamical matrix by introducing an appropriate force field. This gives the free molecule frequencies ω_M which can be checked against the high-frequency internal modes in the solid. Then one assumes that the eigenvectors σ_M of the internal free molecule modes and the internal force field do not change introducing the intermolecular potential. In harmonic approximation the dynamical problem is then reduced to solve the following secular equation:

$$\left| F_{\mu,\ell}^{\nu,m}(\underline{q}) - [\omega^2(\underline{q}) - \omega_M^2] \, \delta_{\mu\nu} \, \delta_{\ell m} \right| = 0 \ . \tag{57}$$

Here μ and ν label molecules in the unit cell, and ℓ and m the 144 normal modes with frequencies $\omega(\underline{q})$. The matrix elements $F(\underline{q})$ of the dynamical matrix contain only external interactions given by the second derivatives of the potential (53).

An important contribution to the gas-solid renormalization of the internal frequencies comes from $\partial^2 \phi / \partial^2 \sigma_M$. The term $\partial^2 \phi / \partial \sigma_{M\nu} \, \partial \sigma_{M,\mu}$ (the derivative with respect to the same internal eigenvector but for different molecules μ and ν) can produce dispersion and a splitting of symmetric and antisymmetric modes without involing a coupling of internal and external modes. This coupling comes in and has to be taken into account, at least for the low-frequency internal modes, by the double derivatives to one internal and one external degree of freedom. Usually the influence of the intermolecular potential on the high-frequency modes is small. Therefore one can restrict the calculation to the low-frequency region.

Unfortunately the free molecule modes of anthracene have not yet been studied experimentally. The reason is probably that the vapor pressure up to temperatures at which the molecule is stable is too low for its investigation. Therefore the low-frequency modes can only be obtained by model calculations which are fitted to the measured (in the solid) high-frequency modes. Such calculations have been performed by EVANS and SCULLY (1964), KRAINOV (1964) and NETO et al. (1966, 1968).

The experimentally obtained dispersion curves, 16 in each symmetry direction (12 external and 4 internal ones), are shown in Fig. 22. For the calculated ones in Fig. 23 we had to assume eigenvectors for the two free molecule internal modes. Each one produces a symmetric and a antisymmetric lattice mode. The symmetry considerations are the same as in naphthalene. From the literature containing calculations and experimental light scattering results we extracted that the lowest internal mode should

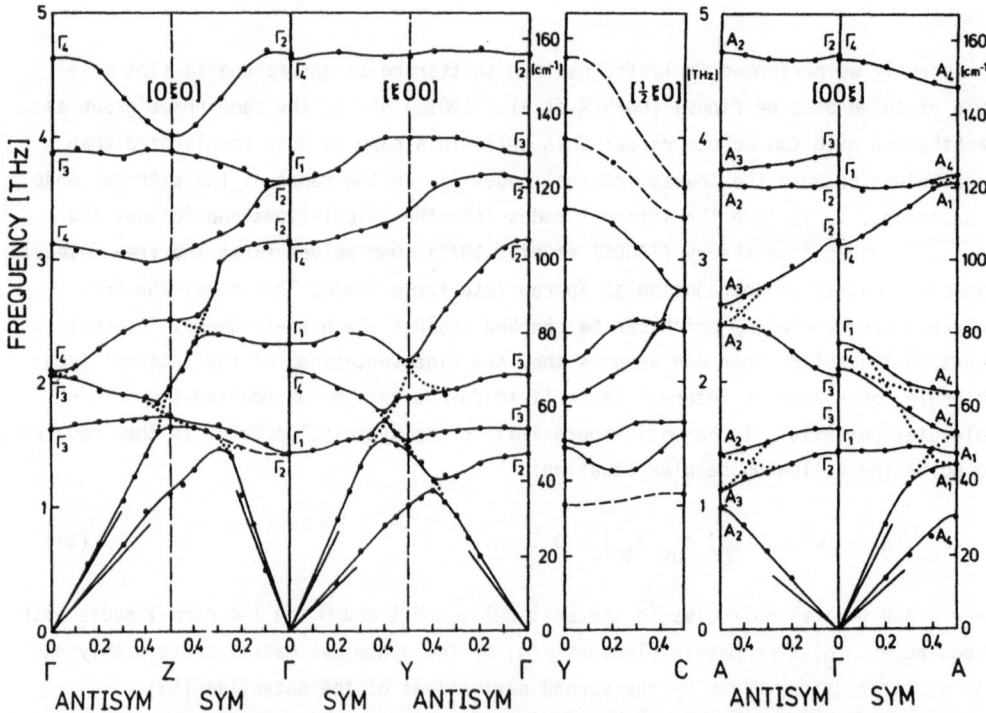

Fig. 22. Measured phonon dispersion curves for the 12 external and the 4 lowest internal modes in anthracene at 12 K for different symmetry directions. The presentation is given in the extended zone scheme such that branches need not cross. In the extended zone scheme the dispersion curves for the [ξ00] and [0ξ0] directions go smoothly through the zone boundary. For an identification of internal modes see Fig.23 (DORNER et al., 1981b)

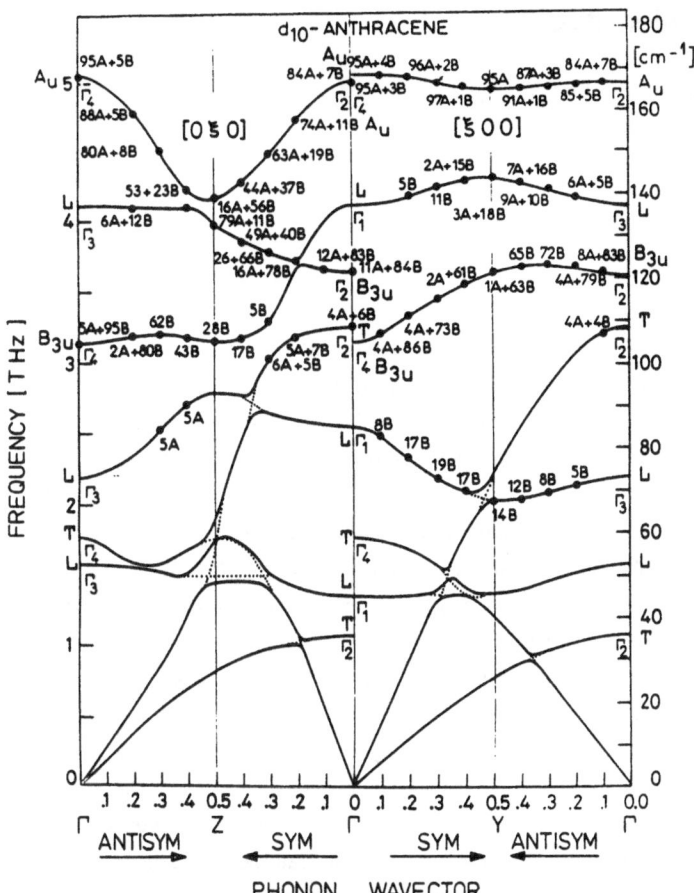

Fig. 23. Calculated phonon dispersion curves for the 12 external and the 4 lowest internal modes in athracene. The internal mode character of the A_u and B_{3u} modes is given in percent, where A stands for A_u and B for B_{3u}. Apparently there is strong mode interaction (DORNER et al., 1981b)

be of B_{3u} type (butterfly mode) and the second of A_u type (twisted around the long axis of the molecule). The calculated frequencies of the internal modes at the zone centre have been adjusted to the measured frequencies by an empirical choice of the corresponding free molecule frequencies ω_M in (57). The overall qualitative agreement with experiment is again very good. An eigenvector determination of the internal modes at the zone centre has been performed (CHAPLOT et al., 1981) and confirmed the B_{3u} and the A_u modes. At the zone centre these modes can only couple to the external translations.

5. Analysis of Phonon Intensities

In the simplest cases the frequency of a phonon is determined by the centre of an intensity distribution as obtained by a const-Q scan. A dispersion curve in a particular symmetry direction is then the connection of frequencies for adjacent values of the phonon wavevector $\underline{q}$ in this direction. Sometimes, if the sample is small or the intensity of the investigated branch is weak, it is not that simple. If the centre of a measured "ugly peak" is badly defined, then it is often preferable to try to measure a neighbouring phonon instead of repeating the first one. It may turn out that the problem is more profound as the eigenvector is changing or as another, unexpected branch comes near in frequency.

In more complicated systems there may be several branches close to each other in frequency. These branches may cross each other if they belong to different irreducible representations or they avoid each other if they belong to the same representation. In the latter case it happens quite often that two branches from the same representation exhibit an anticrossing behaviour in their dispersion curves connected with an exchange of their eigenvectors. In these cases one can very often identify branches by a qualitative analysis of the measured intensity (Sect. 5.2).

If one wants to know the eigenvector of a mode then one needs a quantitative intensity analysis as discussed in Sect. 5.3.

5.1 Electron-Phonon Interaction in Cadmium

In metals the phonon dispersion curves exhibit a fine structure which reflects features of the electron-ion interaction and properties of the itinerant electrons. This effect is principally anharmonic because it contains a scattering of electrons from phonons and vice versa. The existence of a sharp, well-defined Fermi surface results in abrupt changes of phonon frequencies at wavevectors which are related to extremal dimensions of the Fermi surface (Sect. 4.2). Such "kinks", known as Kohn anomalies, have been extensively studied since they were first predicted twenty years ago (KOHN, 1959). The presence of Kohn anomalies in both transition (POWELL et al., 1968; SHAW and MUHLESTEIN, 1971; DUTTON et al., 1972) and non-transition (STEDMAN et al., 1967; BROCKHOUSE et al., 1961; VOSKO et al., 1965) metals is now well established.

BROVMAN and KAGAN (1974) in their treatment of non-transition metals have pre-dicted a hierarchy of phonon anomalies related to the indirect interionic inter-action mediated by the conduction electrons. In addition to the Kohn anomalies which appear when this interaction is central, Brovman and Kagan find further anomalies which reflect the non-central character of the interionic forces. Expressed some-what differently, the Kohn anomalies appear in perturbation calculations of phonon frequencies which are correct to second order in the electron-ion pseudopotential. To find the anomalies predicted by Brovman and Kagan a third-order calculation is required. Recently, a fine structure which appears to confirm the existence of these third-order (or three-particle) anomalies has been found in the phonon spectra of aluminium (RUMYANTSEV et al., 1978).

The experimental task is here to determine dispersion curves with sufficient precision that the group velocity $d\omega/dq$ can be derived. To measure like this, "only" a precise determination of the centre of measured peaks, one needs 1) a sufficiently big sample to count at least 100 neutrons in the peak of the signal, 2) a low and uniform background, 3) a very well aligned TAS, 4) a very well aligned sample, 5) good resolution in q, and 6) possibly a controlled temperature.

Measurements of this kind were performed on cadmium by CHERNYSHOV et al. (1979). Since naturally occurring Cd absorbs neutrons strongly (cross-section 2650 barn), this experiment has been performed with a monocrystal of the isotope ^{110}Cd which had an absorption cross-section of < 2 barn. The sample, grown by the Bridgeman technique, was in the form of a cylinder $\sim$ 20 cm^3 in volume and mosaic spread (FWHM) $\sim$ 16'. To avoid deformation of the sample during the experiment it was mounted in an aluminium sleeve 0.2 mm thick.

The measurements were performed at 77 K. The observed phonons had a width which corresponded to the instrumental resolution. All phonons were fitted to Gaussians modified by the normalization factor $k_F^3 \cot \theta_A$ (Sect. 2.3). From these fits the phonon frequencies and their standard deviations were deduced. Neither the standard deviations nor the differences between repeated scans exceeded 0.002 THz for trans-verse phonons or 0.005 THz for longitudinal modes.

The group velocity $d\omega/dq$ was calculated as $\Delta\omega/\Delta q$, where $\Delta\omega$ and Δq are the diffe-rences of frequency and wavevector between adjacent measured points. Although the absolute accuracy of the measured phonon frequencies (allowing for unknown systema-tic errors) is only 0.02 THz, the relative accuracy is sufficiently good to allow group velocities to be extracted with an error of 5 - 10% (Fig. 24). In this figure we see that the group velocity is not a monotonic function of q, but exhibits ano-malies.

As explained in Sect. 4.2 we expect Kohn anomalies at

$$\frac{1}{2} \left| \underline{q} + \underline{\tau}(h,k,\ell) \right| = k_F \ . \tag{49}$$

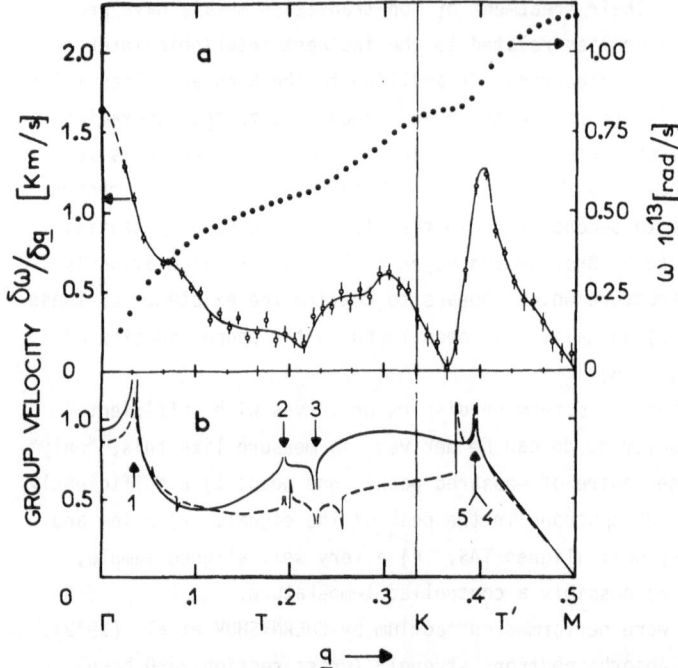

Fig. 24a,b. Kohn anomalies in Cd. (a) Experimental phonon dispersion curve (full circles) and group velocity dω/dq (open circles) for the $T_3 - T_3'$ branch at 77 K. At q = 0 the group velocity obtained from the elastic constants (GARLAND and SILVERMAN, 1960) is represented by a full circle. The line is a guide to the eye. (b) Theoretical predictions of the group velocity for the $T_3 - T_3'$ branch. The full line is calculated in perturbation theory including second-order terms in the potential (46), the broken line including third-order terms in the potential. The numbers refer to predicted anomalies. Only 3 and 4 could be clearly observed. 5 is a very weak (k_F almost perpendicular to the phonon eigenvector) minimum. (CHERNYSHOV et al., 1979)

For a free electron gas k_F would be the radius of the Fermi sphere. Fermi surfaces in conducting materials may be far from spherical due to interaction with the ion lattice. For non-transition metals like Cd, however, the static and dynamic properties can in many cases be well described by (almost) free-electron models. Therefore we begin the analysis of the observed fine structure in the Cd dispersion curves on the basis of results obtained for a spherical Fermi surface.

In Figure 25 the Fermi sphere is drawn superposed on the Brillouin zone scheme for cadmium; the first three zones are filled or partially filled by electron states. According to (49) the positions of expected Kohn anomalies may be found by the following procedure:

I) choose a point corresponding to $\underline{\tau}(hk\ell)/2$ in Fig. 25 (several such points are marked),

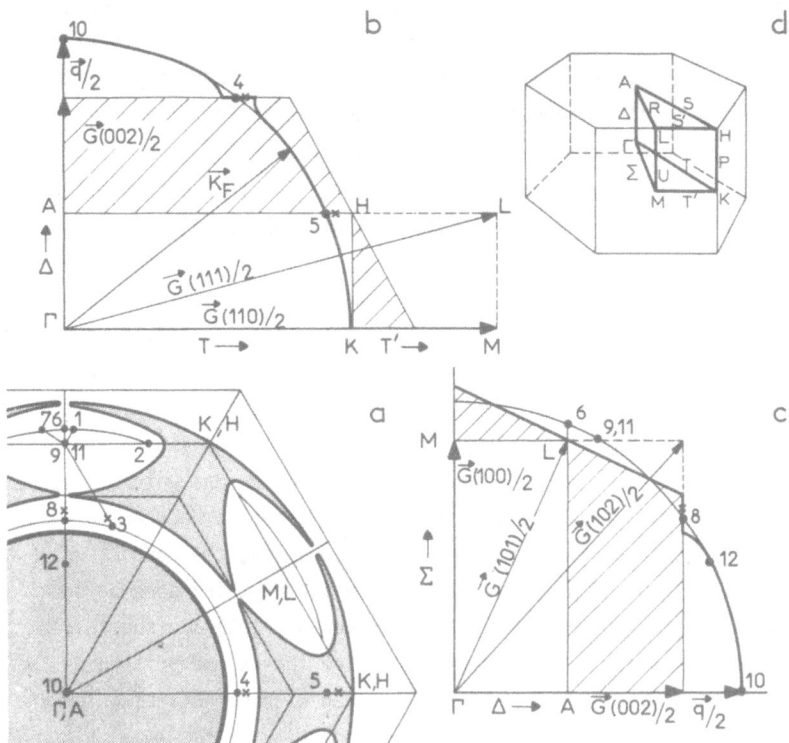

Fig. 25a-d. The Fermi surface of Cd in the scheme of first, second, and third Brillouin zones. (a) Projection on the basal (T - Σ) plane. The Fermi surface for free electrons is shaded while the boundaries of filled bands (boundaries of the second Brillouin zone) are unshaded. (b) and (c) are projections (at cuts) on the Δ - T and Δ - Σ planes, respectively. The hatched regions comprise the second Brillouin zone. The reciprocal space of the phonon system is underlying in a scale 1/2 corresponding to (49). In (a), (b) and (c) the numbers identify predicted anomalies. The full circles are the calculated positions of the anomalies while the crosses are the observed positions. (d) gives the first Brillouin zone with symmetry labels for hcp structures. (CHERNYSHOV et al., 1979)

II) proceed from the chosen point in the direction q (normally a symmetry direction) until the Fermi surface is encountered,

III) if the distance between the chosen (starting) point and this Fermi surface intersection is $|q/2|$, the Kohn anomaly will be found at wavevector q.

It is to be noted that for spherical Fermi surfaces a Kohn anomaly can only occur for a phonon whose polarization vector has a component parallel to the corresponding k_F. Thus, for example, vectors $\tau(h,k,0)$ cannot contribute in (49) to anomalies in the T_3, T_3' and Σ_3 modes which propagate in the basal plane and are polarized (by symmetry) along [001].

This restriction is lifted for non-spherical Fermi surfaces where two types of electron transitions contribute to Kohn anomalies; those for which the wavevectors of the initial and final electron states are related by inversion symmetry and those for which no such symmetry exists. The latter appear for transitions, for example, between parallel flat parts of the Fermi surface. KAGAN et al. (1981) have shown that this second type of transition has a relatively small effect on phonon frequencies in Cd. Therefore only transitions between diametrically opposite electron states were considered.

The positions of the observed anomalies and the absence of some anomalies, expected for a spherical Fermi surface, confirm the deviations from sphericity (Fig. 25) as already proposed by VISSCHER and FALICOV (1972), STARK and FALICOV (1967), GAIDUKOV (1970), and references cited by these authors. The lattice potential strongly distorts the Fermi surface from sphericity by producing an energy gap in the electron states at the parts of the boundary of the second Brillouin zone which is dashed in Fig. 25b and c. Thus the number of conduction electron states is reduced to about one half of its value for a spherical Fermi surface. For example, the conduction electron states, which for a spherical Fermi surface lie in the third zone beyond the zone boundary which bisects the reciprocal lattice vector [101] are suppressed in the real system (i.e. transferred to other areas inside the second Brillouin zone). All unshaded areas in Fig. 25a correspond to filled electron states below an energy gap along the boundary of the second Brillouin zone.

As Cd has two conduction (perhaps better, valence) electrons per atom and two atoms per unit cell, there are 4 electrons per unit cell in real space. The volume of their states in reciprocal space is equal to the volume of two Brillouin zones, because two electrons (one spin up, one spin down) per unit cell could just fill the first Brillouin zone. This means that for Cd there are as many empty states in the second Brillouin zone as occupied states in the third Brillouin zone. The observed and unobserved (for white areas in Fig. 25a) anomalies confirm the expectations as far as their positions are concerned, but the shapes and strengths differ largely from the calculated predictions.

Calculations were carried out using a local, Abarenkov - Heine model potential (43) (ABARENKOV and HEINE, 1965) fitted to the Cd phonon spectrum by DORNER et al., (1981c). The band-structure contribution to the dynamical matrix was calculated in lowest order. This gives a contribution to the frequencies as given by (50) in Sect. 4.2. The phonon dispersion curve should exhibit a step upwards if one follows the phonon wavevector through the Fermi surface from inside, where the electron susceptibility $\chi(q)$ is finite, to outside, where $\chi(q)$ goes to zero, and a step downwards if one goes from outside to inside the Fermi surface. The calculated group velocity in turn exhibits simple maxima or simple minima (Fig. 24). The observed anomaly No. 4 (twofold degenerate) shows a strong maximum at the expected position, but in addition a deep unexpected minimum for slightly smaller q values.

The strength of the observed anomalies is much bigger than expected from calculation. This difference may have two reasons: I) the electron susceptibility in Cd
is stronger than calculated, or II) the screened model potential which enters the
electron phonon coupling constant is not properly adapted.

Anomalies 4 and 4' (degenerate on the symmetry direction) could be separated by
measuring the dispersion curve along parallel paths in reciprocal space. Fig. 26
shows the reciprocal space with the Brillouin zone around the origin. This figure
is complementary to Fig. 25 in which the electron wavevectors are drawn as starting
from the origin of reciprocal space. The phonon wavevectors q were drawn in scale 1/2
and were considered to have any reciprocal lattice vector as origin. In Fig. 26 the
presentation is the inverse; the phonon wavevectors are considered as starting from
the origin of reciprocal space and the electron wavevectors are drawn in scale 2
and as originating from different reciprocal lattice vectors. Under the simplified

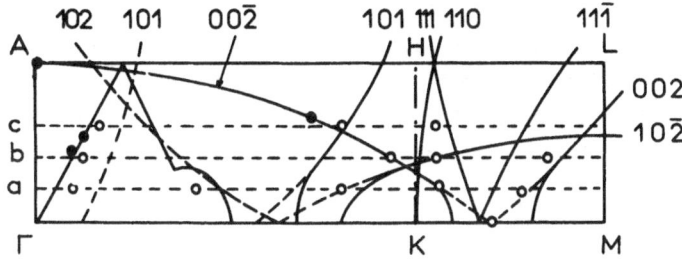

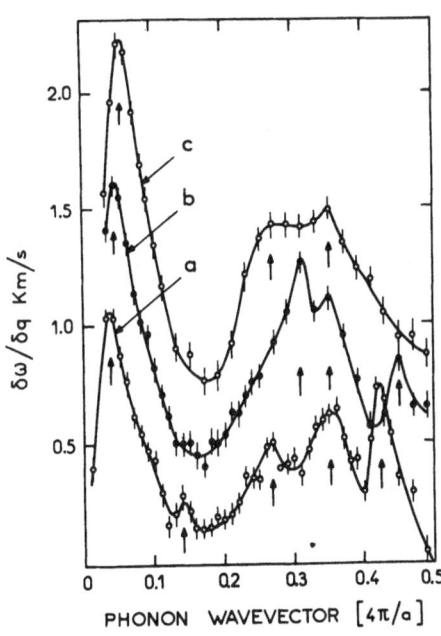

Fig. 26. The Γ-K-M-L-H-A plane of the reciprocal space of the phonon system. Parts
of circles with radius $2k_F$, (49), around
various reciprocal lattice points indicate
possible Kohn anomalies. Full lines represent the real Fermi surface and dashed
lines the hypothetical spherical Fermi
surface in areas where the real one is
flat, as for example orinating from (101).
At (o) anomalies in dω/dq could be clearly
observed for scans along a,b,c (Fig. 27).
At (•) anomalies were detected in scans
perpendicular to a-c. (CHERNYSHOV et al.,
1981)

Fig. 27. Phonon group velocities dω/dq
along paths a,b,c in reciprocal space as
indicated in Fig. 26. Anomalies are marked
by arrows. They correspond to open dots
in Fig. 26. (CHERNYSHOV et al., 1981)

assumption that the Fermi surface is a sphere, Kohn anomalies can be expected in the phonon Brillouin zone around the origin on spherical surfaces with radius $2k_F$ centred at surrounding reciprocal lattice points. From Fig. 26 it is evident that anomalies 4 and 4' corresponding to (002) and (00$\bar{2}$) coincide for the T - T' symmetry direction. The dispersion curves parallel to the T - T' direction but displaced by q_c in the c direction have been measured (Fig. 26, 27). Anomalies 4 and 4' are now separated. Both are expected to exhibit maxima as q runs from inside the corresponding $2k_F$ spheres to outside. The maxima are clearly observed where they were expected (see arrows in Fig. 27). As mentioned before, unexplained minima are observed in each case for slightly smaller q vectors.

In Fig. 24 a second theoretical curve (broken curve) is presented, the calculation for which includes non-pair interionic interactions mediated by the conduction electrons (BROVMAN and KAGAN 1974). In addition to the Kohn anomalies found in the second-order calculation (full curve), additional anomalies appear which arise from third-order terms in the perturbation expansion of the electron energy in powers of the model potential. Several such anomalies appear to have a magnitude similar to those of observable Kohn anomalies. However, we have not been able to unambiguously identify any three-particle anomalies in our data. In the T and Σ directions the data do not permit a positive identification. For the T' direction the calculated three-particle anomaly is probably too close to the strong observed Kohn anomaly to be separable from the latter. Finally, in the Δ direction, the predicted, strong three-particle anomaly is certainly absent.

A three-particle anomaly is expected to occur at wavevector $\underline{q}$ if the vertices of a triangle whose sides are $\underline{\tau}_1 + \underline{q}$, $\underline{\tau}_2 + \underline{q}$, $\underline{\tau}_3 = \underline{\tau}_1 + \underline{\tau}_2$ ($\underline{\tau}_1, \underline{\tau}_2, \underline{\tau}_3$ are reciprocal lattice vectors) lie on the Fermi surface. The detailed shape of the Fermi surface is thus important in determining both the positions and magnitudes of these anomalies. Given that the geometrical condition described above is satisfied, the magnitude of the three-particle anomaly depends on the product of the Fourier components of the model potential with wavevectors $\underline{\tau}_1 + \underline{q}$, $\underline{\tau}_2 + \underline{q}$ and $\underline{\tau}_3$. The strength of the anomaly is thus sensitive to the detailed shape of the model potential rather than to its value at a single wavevector. Relatively small changes of the model potential, which have little overall effect on phonon frequencies could cause a drastic change in the magnitudes of three- particle anomalies and explain the failure to observe such anomalies. Finally, it is possible that the use of a local model potential, fitted to phonon frequencies, is not adequate for a discussion of three-particle anomalies in cadmium. Further theoretical efforts are needed to produce a model which is able to describe the experimental observations.

In this section the intensity of the observed phonons was not analysed. Only statistical accuracy was provided to determine peak positions sufficiently well.

5.2 Anticrossing of Dispersion Branches and Exchange of Eigenvectors in Naphthalene

In this section we consider peak positions and peak intensities at the same time as following one or two dispersion curves. Let us consider 2 branches, say, one translational optic and one librational (rotational oscillation of a rigid molecule), which is always optic. These two modes shall belong to different representations in the zone centre at Γ and shall be pure translations and pure librations at Γ. Following their dispersion along a symmetry direction they may belong to the same representation for this direction and at the zone boundary to different ones.

Let us assume that at Γ the frequency of the libration ω_{lib} is below the frequency of the translation ω_{tr} and that it is the opposite at the zone boundary. Then there is a crossing region in q where the two modes exchange their eigenvectors, i.e. in this region there exist two modes of mixed motion of libration and translation with different frequencies. These modes repell each other as can be calculated in harmonic approximation by a coupling matrix

$$
\begin{pmatrix} \omega_{lib}^2(q) & \Delta(q) \\ \Delta(q) & \omega_{tr}^2(q) \end{pmatrix}
\tag{58}
$$

with the coupling parameter $\Delta(q)$. The corresponding eigenvectors are pure libration and pure translation. After diagonalization we find two frequencies

$$
\omega_{1,2}^2 = \frac{\omega_{lib}^2(q) + \omega_{tr}^2(q)}{2} \pm \frac{1}{2}\sqrt{[\omega_{lib}^2(q) - \omega_{tr}^2(q)]^2 + 4\Delta^2(q)} \; .
\tag{59}
$$

The eigenvectors corresponding to ω_1 and ω_2 are mixed as already mentioned. The coupling parameter may be strong or weak and more or less extended in q. The task is now to observe these two branches experimentally.

For complicated systems with many atoms (like naphthalene) it is hopeless to start experimental work without a model by which eigenvectors and structure factors G_j (19) can be predicted. Quite often the symmetry requirements are so strong for particular points, like the zone centre Γ, that the predicted eigenvectors are quite correct although the predicted frequencies are far off. The calculation of $|G_j|^2$ has to be performed for all Brillouin zones accessible in the experiment. From the calculation one selects a zone, where $|G_j|^2$ for one mode is very large compared to all others in the neighbourhood in energy (in our example of only two modes, one selects two zones, one for the libration and one for the translation).

The experiment starts at Γ to test the agreement between calculated and measured intensities. Quite often one uses information from Raman or IR results to know the frequencies ω_{lib} and ω_{tr} at Γ and thus to correlate the maxima, as obtained in an initial inelastic neutron experiment, to the types of modes. After identification of the modes at Γ one proceeds to follow each one into the Brillouin zone along a symmetry direction, performing constant-Q scans at increasing values of $|q|$.

Let us assume that it was possible to follow ω_{lib} and ω_{tr} in different zones from Γ to the zone boundary. Then we only know that we have followed a particular eigenvector but it is not conclusive that we have followed a dispersion branch, because we assumed that both modes belong to the same irreducible representation. We have to plot the observed values for both frequencies versus q and then connect the points by two continuous lines which do not cross. These dispersion curves will start at Γ as libration and arrive at the zone boundary as translation.

If $\Delta(q)$ is small and not extended in q very much, one can easily miss an anticrossing or eigenvector exchange mecanism if one follows only one type of eigenvector. This happened for ZnO, where OSTHELLER et al. (1968) published dispersion branches which are not correct. THOMA et al. (1974) observed later an optic mode which has an anticrossing effect with one acoustic branch.

The results obtained for the dispersion curves of d_8-naphthalene at 6 K in five directions are given in Fig. 13 (NATKANIEC et al., 1980). Vertical and horizontal bars indicate errors for the constant-Q and constant-E mode of operation, respectively.

In the structure of naphthalene there are at most 4 symmetry operations generally leading to 4 non-degenerate irreducible representations or to one doubly degenerate irreducible representation. At Γ (and at A) the eigenvectors are either pure librations or pure translations: Γ_1 and Γ_3 the symmetric and antisymmetric (in the screw diad and in the glide plane) librations, and Γ_2 and Γ_4, the symmetric and antisymmetric (in the screw diad and opposite in the glide plane) translations. At point C(1/2 1/2 0) there are 4 non-degenerate irreducible representations. But time reversal symmetry leads to two doubly degenerate reducible representations: C_1 pure librations and C_2 pure translations. Along $|1/2\ \xi\ 0|$ the one doubly degenerate representation is reducible as a consequence of time-reversal symmetry. In the symmetry directions Δ and θ the full and dashed lines indicate the symmetric and antisymmetric representations, respectively. At the zone centre, there are nine optic modes $3\Gamma_1 + 2\Gamma_2 + 3\Gamma_3 + 1\Gamma_4$. The corresponding labels used in light scattering are: $\Gamma_1 - A_g$, $\Gamma_2 - A_u$, $\Gamma_3 - B_g$, $\Gamma_4 - B_u$; Γ_1 and Γ_3 are observed in Raman scattering by ITO et al. (1968). Both Γ_2 and Γ_4 ar IR active but no data on d_8-naphthalene at 6 K have been published. The optic data are in very good agreement with the neutron results.

The crossings of eigenvectors which occur at the anticrossing of branches within the same representation are indicated by dotted lines in Fig. 13. Most of them were verified experimentally, as can be seen in Fig. 28. In the centre of the zone (410) the symmetric libration Γ_1 (2.52 THz) has a much higher intensity than the antisymmetric Γ_3 (2.38 THz) libration. In the neighbouring zone (420) it is just the opposite. This effect was predicted by calculations and experimentally observed. As naphthalene has a non-symmorphic space group, one can expect that the two librational branches emerging from these Γ_1 and Γ_3 points into the Δ direction can be

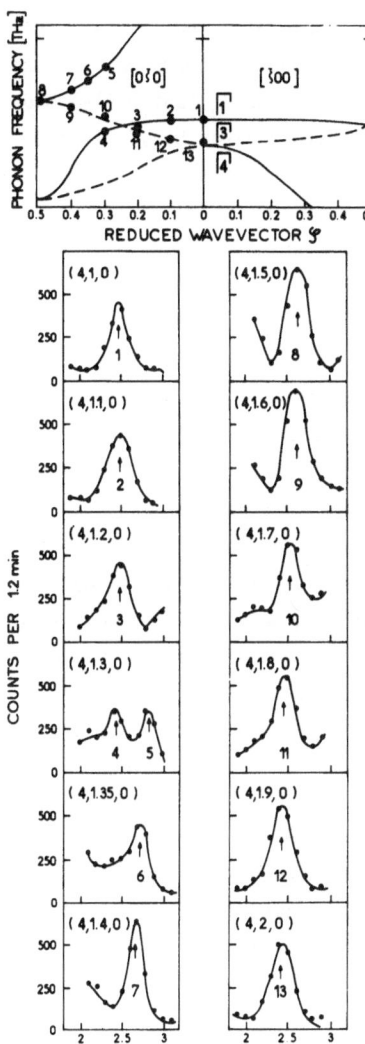

Fig. 28. Eigenvector crossing effect in naph-thalene for the [0ξ0] direction. The lower part displays several constant-Q scans at different momentum transfers running from (410) to (420). The plots represent rough data obtained with fixed outgoing energy (3.55 THz). Observed maxima are numbered and plotted in the upper part of the Figure onto the corresponding dispersion curves. (NATKANIEC et al., 1980)

observed along a path connecting (410) and (420). A series of phonons of this type is shown in Fig. 28. The anticrossing effect appears at (4 1.3 0) where two maxima of reduced intensity were observed because the librational character of the eigen-vector is partly on the branch emerging from Γ_2 (3.20 THz). The different phonons on different branches are identified in this way. The separation in energy of Γ_1 and Γ_3 is not obtained by resolution but by visibility.

This technique to determine two branches connected at the zone boundary by fol-lowing them from one Brillouin zone into a neighbouring one is particularly appro-priate for non-symmorphic space groups. It is often called working in the extended zone scheme (Sect. 3.3).

The four lowest internal modes were measured in the three principal symmetry directions and at the C point (1/2 1/2 0) (Fig. 13). The probable dispersion of these modes in the other two directions is shown by dotted lines. The B_u (butterfly) mode (in notation of the free molecule point group D_{2h}) exhibits a clear Davydov splitting in the zone centre (170/182 cm^{-1}) which agrees well with the splitting observed in the IR spectrum at room temperature (164/178 cm^{-1}; BREE and KYDD, 1970). This splitting agrees quantitatively with results from IINS experiments (BOKHENKOV et al., 1976). The splitting of the A_u mode remains within the error bars.

The observation of Davydov splitting in internal modes does not necessarily mean that there is a coupling of internal and external modes as explained in Sect. 4.3. But coupling is highly probable because the frequencies are so close to each other. Anyway, the splitting is sufficiently small that the assumption of rigid molecules for the calculation of the external modes should be a good approximation. Davydov splitting and mode repulsion are perfectly treated in harmonic approximation.

The qualitative picture of the calculated curves is already quite similar to the true result. This is astonishing if one realizes that the parameters of the model are not fitted to the measured frequencies but derived from macroscopic observations. Yet one may say that there is certainly some truth in the atom-atom potential, as the calculated frequencies come from an average over many atom-atom interactions at many different distances. Therefore it may be less surprising that the qualitative picture turns out to resemble the truth. In this context, it is interesting that POWELL et al. (1978) found two different and equally satisfactory sets of parameters to describe the frequencies of benzene. To look into this problem in more detail, an eigenvector determination for 5 zone boundary modes of naphthalene has been performed (PAWLEY et al., 1979). It was found that the set of data which describes the frequencies qualitatively well reproduces the eigenvectors reasonably well.

5.3 Eigenvector Determination

In Sect. 5.2 we described how a qualitative consideration of measured phonon intensities allows distinguishing one dispersion branch from another. A quantitative analysis of measured phonon intensities goes much further as it allows determining the mode eigenvector $\underline{\sigma}$ (19). An eigenvector determination might be called a dynamical structure determination. From a normal structure determination one obtains the equilibrium positions $\underline{d}$ (19) of the atoms in the unit cell by analysing the intensities of many Bragg reflections.

To perform an eigenvector determination one has to already know the structure. The task is then to obtain and analyse phonon intensities for many different momentum transfers $\underline{Q}$, always measuring the same phonon. This means the phonon wavevector $\underline{q}$ has to be always identical.

An eigenvector determination is much more time consuming than a structure determination because phonon intensities are about 3 orders of magnitude weaker than
Bragg reflexes. Much care has to be taken to avoid spurious effects such as an artificial increase of intensity by multiple scattering or an artificial decrease by
simultaneous Bragg reflections. The knowledge of one or several eigenvectors is required for several reasons as discussed in Sect. 5.3.1.

5.3.1 Eigenvectors and Lattice Dynamical Models

As already explained in Chap. 4 lattice dynamical models provide first hand a para-
metrisized description of the experimentally observed phonon dispersion curves. The
physical microscopic relevance of the force constants in a Born-von Karman model is
highly questionable. LEIGH et al. (1971) and COCHRAN (1971) showed that the dispersion curves of a given substance can be equally well described by different sets of
force constants as far as the frequencies are concerned. A decision which model is
closer to reality can only be made on the grounds of experimentally determined eigenvectors.

In the case of zinc this problem was investigated in detail by CHESSER and AXE
(1974). At the time most of the dispersion curves had been measured elsewhere and
two models were proposed to describe the data: a Born-von Karman model by DeWAMES
et al. (1965) and a pseudopotential model by GILAT et al. (1969). Zinc has a hcp
structure and thus two atoms per unit cell; in other words, it consists of two sublattices. For symmetry directions in a monatomic hcp structure the atom polariza-
tions in different modes are determined by symmetry. The only free parameter in the
eigenvector is the sign of the phase between the two sublattices. CHESSER and AXE
(1974) investigated this phase factor for the longitudinal modes in [010] direction.
The pseudopotential calculations agreed with the experimental observation, while the
proposed set of force constants in the Born-von Karman model predicted an opposite
sign of the phase. Chesser and Axe were able to define another set of force constants
in the same Born-von Karman model to get the sign of the phase right and a comparably
good description of the frequencies. Nevertheless, a comparison between the models
revealed that the pseudopotential calculation by GILAT et al. (1969) most closely
approximates the data. From Sect. 4.2 one would have even expected that a pseudopotential model including the electronic part in the bonding has advantages in describing the lattice dynamics of a metal.

5.3.2 Exchange of the Transverse Mode Eigenvectors in AgBr at the L point

Model calculations for AgBr (FISCHER et al., 1972) (Sect. 4.1) suggested significant qualitative differences as compared to the corresponding alkali halides. The differences are due to the deformability of the Ag^+ ion d shell and its partially covalent coupling to the neighbours. The most striking feature predicted theoretically is the inversion of the transverse optic and acoustic mode at the L point. It results from a crossing of the two branches that have the same symmetry.

Although phonon-assisted exciton recombination (Von der OSTEN and WEBER, 1974) and second-order Raman scattering (Von der OSTEN, 1974) supported the validity of the model, it was unambiguously proved in an eigenvector determination at 85 K by Von der OSTEN and DORNER (1975).

At the L point of diatomic fcc structure the eigenvectors have the particular behaviour, that in each mode only one sublattice vibrates, while the other one is at rest. In consequence the eigenvector of a L point mode contains only the amplitude of one ion. Thus the structure factor (19) in transvers geometry $q \perp Q$ reduces to

$$G_j(g,Q)^2 = b_j^2 \frac{Q^2}{M_j} e^{-2Wj} . \tag{60}$$

Here j indicates the ion vibrating in mode j.

The fact that in each mode only one sublattice is moving also implies that always both the optic and the acoustic modes are visible. Therefore one measurement at one L point is sufficient to determine which ion is oscillating at which frequency. But to determine the intensities satisfactorily one needs good resolution to separate the two contributions.

It was an experimental problem to obtain the necessary resolution in the presence of the strongly absorbing Ag (63 barns at 6 THz neutron energy). The resolution is most easily improved by reducing the neutron energy but at the same time increasing the absorption proportional to $1/k_I$; thus the useful sample volume is reduced.

The experimentally observed intensity distribution at the L point (2.5, 1.5, 1.5) is presented in Fig. 29. It was fitted by a constant background and two Gaussians modified by the resolution normalization (Sect. 2.1). The best fit is shown by the full line, the two phonons having frequencies of 1.54 and 1.93 THz, respectively. Since the necessary resolution corrections were included in the fit program, one could derive the integrated intensity ratio to be

$$(I_{TA}/I_{TO})_{exp} = 2.15 \pm 0.4 . \tag{61}$$

A theoretical value can easily be calculated from (60). The difference in the Debye-Waller factor for Ag^+ and Br^- could be neglected at the low temperature and the small momentum transfer.

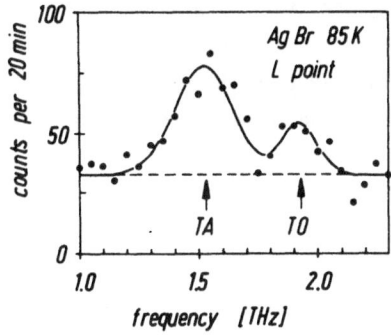

Fig. 29. The transverse TA and TO phonons at the L point in AgBr as measured in a constant-Q scan. A constant background (----) is subtracted from the data which were fitted by two Gaussians modified by the resolution normalization (16) (Von der OSTEN and DORNER, 1975)

According to the model by FISCHER et al. (1972) the lighter Br$^-$ ion should vibrate with the lower frequency in the TA mode and the heavier Ag$^+$ ion with the higher frequency in the TO mode. For this case the calculated ratio was

$$(I_{TA}/I_{TO})_{calc} = 2.38 \qquad (62)$$

which is in good agreement with the experimental result. In the opposite case one would obtain a ratio of 0.85 which is far off the value found experimentally.

The fact that in this case the lighter ion vibrates with a lower frequency than the heavier one is unusual and implies that the forces active in the two modes are different. If the forces would be the same the frequencies should be

$$\frac{\omega_1^2}{\omega_2^2} = \frac{M_2}{M_1} . \qquad (63)$$

In the case of AgBr the quadrupolar deformability of the 4-d electron shell of the Ag$^+$ is the reason for the different forces active in the different modes.

5.3.3 Eigenvector Exchange of Two Modes with Varying Temperature in Quartz

Quartz, a material which has many important technical applications, is not yet fully understood in its structural and dynamical properties. This is partly related to the fact that the unit cell contains 9 atoms. This makes it a complex system. It exhibits a phase transformation from the trigonal α- to the hexagonal β-phase at 846 K. In both phases quartz consists of a framework of nearly regular SiO_4 - tetrahedra linked together by common corners, showing some characteristic structural subunits like chains and spirals. The main difference of the α- to the β-phase is a rotation of the SiO_4 tetrahedra around an axis in the basal plane (16° at room temperature). This axis contains a symmetry element in the β-phase (twofold axis) which is lost in the α-phase, and the rotation around it decreases the bond angle Si-O-Si. In the β-phase the Si-O-Si angle has a maximum with respect to the above twofold axis. A geometrical model for this transition based on rigid SiO_4 tetrahedra was proposed by GRIMM and DORNER (1975).

Numerous optical investigations (for a review see SCOTT, 1974) and a neutron scattering experiment by AXE and SHIRANE (1970) revealed the instability of a normal mode (soft mode) at the Γ point responsible for the α-β transition, although this mode could not be detected directly near the transition temperature due to enhanced damping. In addition to the Γ point anomaly, experimental investigations by X rays (ARNOLD, 1965 and 1976) and neutrons (BAUER et al., 1971; DORNER et al., 1974) showed an intense diffuse scattering along the whole Γ-M direction, being of inelastic origin.

Approaching the α-β transition a softening of the two lowest lying acoustic branches was observed (Fig. 30). Particularly the zone-boundary modes (M point)

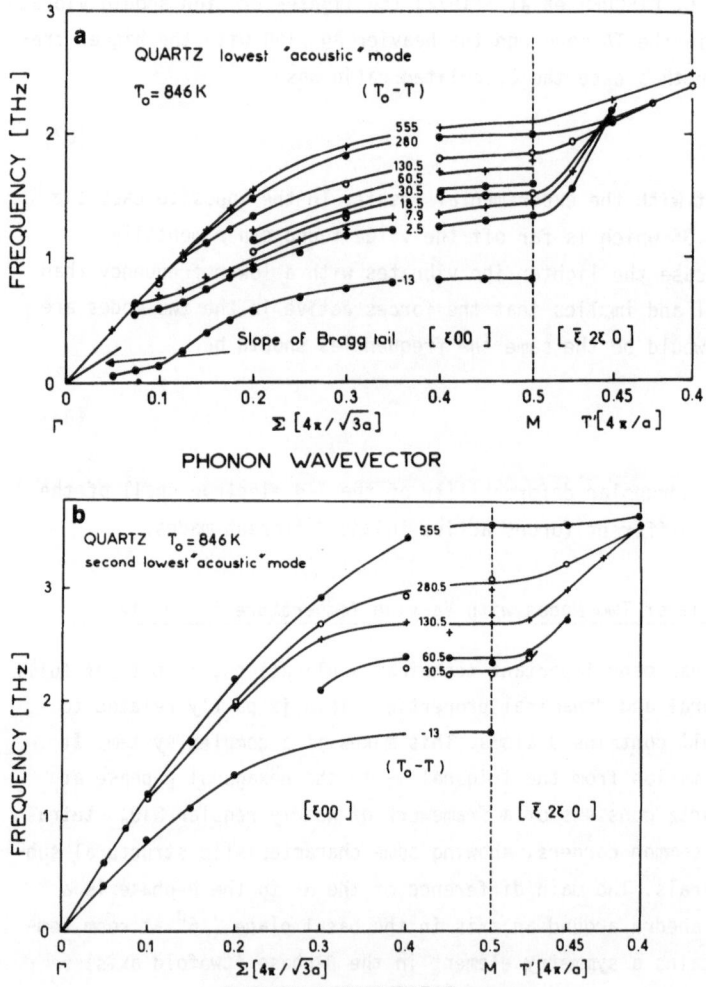

Fig. 30a,b. Phonon dispersion curves in quartz at different temperatures. T_0 is α-β phase transformation temperature. Note that the T' direction is perpendicular to the Γ-M direction. (BOYSEN et al., 1980) (a) The lowest and (b) the second lowest "acoustic" mode

soften appreciably. From the temperature dependence of these modes - the upper mode
softens rapidly at low temperatures and the lower mode softens rapidly at high temperatures - one suspects that one strongly temperature-dependent mode crosses a
practically stable mode (Fig. 31). This behaviour is proved experimentally as de-

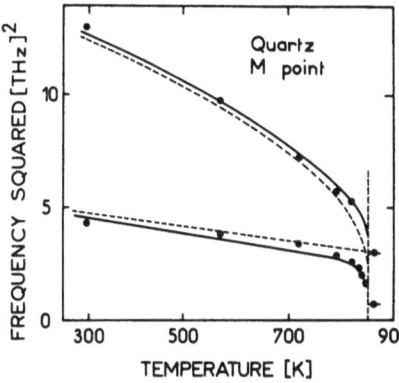

Fig. 31. The temperature dependence of the
lowest modes at the M point in quartz. Low-
frequency series from Fig. 30a and high-frequency series from Fig. 30b. The solid lines
are a fit of (70) and the dashed lines represent the temperature dependence of the uncoupled frequencies (66, 68). (BOYSEN et al.,
1980)

scribed in the following: To reveal the microscopic origin of the softening of modes
along Γ-M, particularly to find a connection with the true Γ-point soft mode (SM),
a dynamic structure investigation was started. By means of an eigenvector determination of the interesting temperature-dependent M-point modes having real eigenvectors,
one obtained an insight into the dynamic behaviour of the SiO_4 framework during the
α-β phase transformation (BOYSEN et al., 1980).

Data were collected at 3 different temperatures: room temperature (296 K), 573 K,
and 859 K which is 13 K above the α-β phase transformation. The low-frequency mode
was observed at all three temperatures, while the second lowest mode at room temperature was indistinguishable from neighbouring modes.

As absolute intensities had to be determined, two data sets were always measured
with two different wavevectors k_I of the incoming neutrons, 4.6 and 4.9 $Å^{-1}$. Spurious effects which may artificially increase or decrease a particular phonon intensity depend strongly on k (Sect. 5.3). Comparing the relative intensities of the
two data sets dubious scans were eliminated.

The observed intensities were least squares fitted by Gaussians which had been
modified by the resolution normalization (Sect. 2.1). The positions (frequencies)
and the widths (mainly instrumental resolution) were obtained at each temperature
by fitting strong phonons of each mode first. As frequency and width of a mode with
horizontal dispersion which is necessarily the case at the M point have to be independent of Q, frequency and width could be kept fixed when fitting the remaining

scans for one data set. These integrated intensities are directly proportional to the squared inelastic structure factor (19).

The unit cells of α- and β-quartz contain 9 atoms. Thus the eigenvector has 27 components which have to be determined. This number of free parameters may be reduced considerably by applying group theory. In table 2 the symmetry elements of the wave-

Table 2. Group theoretical representations of the little group at the M point in α- and β-quartz. C_2 and D_2 are the corresponding point groups, C_2^α are twofold rotations around axes α, n is the number of modes belonging to each representation, n_T the number of modes which leave the SiO_4 tetrahedra rigid, and n_s the mode which leaves the SiO_4 tetrahedra rigid and keeps the bond angle Si-O-Si constant

α-quartz:

C_2	E	C_2^x	n
m_1	1	1	13
m_2	1	-1	14

β-quartz:

D_2	E	C_2^x	C_2^y	C_2^z	n	n_T	n_s
M_1	1	1	1	1	6	0	0
M_2	1	-1	-1	1	8	2	1
M_3	1	-1	1	-1	6	0	0
M_4	1	1	-1	-1	7	1	0

vector group at the M point are given together with the representations in the α- and β-phases. The compatibility relations between the representations in the α- and β-phase are

$$m_1 \Big\langle \begin{matrix} M_1 \\ M_4 \end{matrix} \qquad m_2 \Big\langle \begin{matrix} M_2 \\ M_3 \end{matrix} \,.$$

Let n be the number of modes in each representation. Then n - 1 is the number of free components of the eigenvector for a particular mode (normalizing condition). In the β-phase the number of modes (n_T) that leave the SiO_4 tetrahedra rigid were calculated as well as those (n_s) which in addition keep the Si-Si distance (the Si-O-Si bond angle) constant. As there is only one mode with rigid tetrahedra and constant

Si-Si distance (in representation M_2), the corresponding eigenvector could be calculated directly from structural parameters. Then by applying the orthogonality condition the remaining rigid tetrahedra mode in M_2 was calculated as well. It

It was not immediately evident which representation the observed modes belong to.
From analogy to observed dispersion curves in other hexagonal and trigonal systems
the lowest mode at room temperature was interpreted as the lowest acoustic mode predominantly polarized in c direction. Therefore it had to be m_2. For the second mode
it was the quality of the fit which significantly favoured the m_2 representation.
In the β-phase it was intriging that 2 rigid tetrahedron modes appear in M_2 and none
in M_3. It was expected that low-frequency modes do not distort the hard SiO_4 cage.
Nevertheless it was tried to fit the observed integrated intensities, too, by the
parameters in the other representations. Finally the quality of the fit was decisive
to attribute both modes to the M_2 representation. Typical extinctions for M_2 were
observed as well.

The free parameters of the m_2 representation in α and of the M_2 representation in
β have been fitted to the observed intensities. The observed and calculated intensities are given in Table 3. The fit routine was started from several very different
sets of "first-guess" values for the free parameters. The calculations always con-
verged to the same eigenvector. The orthogonality condition was not included in the
fit. The two eigenvectors of the modes had an angle of 96° at 573 K and of 104° at
859 K. The deviation is of course unphysical and reflects the experimental uncer-
tainties. The obtained eigenvectors of the upper mode at 573 K and of the lower mode
at 296 K turned out to be very close to those calculated by BARRON et al. (1976)
with a lattice dynamical model taking the structure of quartz at room temperature.

For a detailed analysis one best starts from the eigenvectors $\underline{\sigma}^{\ell\beta}$ of the low-
frequency mode and $\underline{\sigma}^{u\beta}$ of the upper mode in the β-phase. As mentioned earlier, the
eigenvector of the rigid tetrahedra mode with fixed Si-O-Si bond angle $\underline{\sigma}_1$ in M_2 can
be calculated completely, and then by orthogonality the second rigid tetrahedra mode
$\underline{\sigma}_2$ can be calculated as well. These calculated eigenvectors have been taken as a
basis to interpret the eigenvectors obtained experimentally as linear combinations
of these two basis vectors,

$$\underline{\sigma}^{\ell\beta} = \underline{\sigma}_1 \cos \alpha + \underline{\sigma}_2 \sin \alpha \,,$$

$$\underline{\sigma}^{u\beta} = -\underline{\sigma}_1 \sin \alpha + \underline{\sigma}_2 \cos \alpha \,.$$

(64)

As $\underline{\sigma}^{\ell\beta}$ and $\underline{\sigma}^{u\beta}$ are not perfectly orthogonal we obtain different α values. The $\underline{\sigma}^{u\beta}$
turns out to be close to $\underline{\sigma}_2$ ($\alpha \approx 7^{\circ}$). The lowest mode is similar to $\underline{\sigma}_1$ ($\alpha \approx 13^{\circ}$).
The $\ell\beta$ mode in β is apparently well described by the rigid tetrahedra mode, which
leaves the Si-O-Si bond angle fixed.

Since it is known that the soft mode at Γ in the β-phase is also of this type,
it is straightforward speculation that the low-frequency "optic" mode in Σ direction

Table 3. Observed and calculated integrated phonon intensities of the two lowest modes in quartz at the M point at different temperatures. (ℓ) and (u) indicate the lower and upper mode, respectively. The R factors of this eigenvector determination are given as well

h	k	ℓ	296 K (ℓ)		573 K (ℓ)		573 K (u)		859 K (ℓ)		859 K (u)	
			I_{calc}	I_{obs}	I_{calc}	I_{obs}	I_{calc}	I_{obs}	I_{calc}	I_{obs}	I_{calc}	I_{obs}
2.5	0	0	10	8(3)	3	3(3)	0	5(5)	0	0(8)	0	0(8)
3.5	0	0	315	302(32)	187	180(14)	23	30(11)	0	0(5)	0	0(5)
4.5	0	0	14	12(5)	13	14(3)	0	0(3)	0	0(5)	0	0(5)
0.5	2	0	15	17(3)	12	15(3)	1	0(1)	30	33(6)	0	4(5)
0.5	3	0	18	20(4)	10	11(3)	23	27(6)	187	149(22)	10	18(10)
0.5	4	0	60	63(5)	14	13(4)	27	25(7)	163	157(15)	2	18(8)
0.5	5	0	69	64(10)	58	40(7)	26	16(8)	150	141(11)	14	4(8)
0.5	-3	0	85	89(7)	79	81(6)	12	11(5)	118	127(17)	90	89(12)
0.5	-4	0	60	63(6)	79	79(5)	31	32(5)	593	601(50)	45	44(10)
0.5	-5	0	13	13(3)	10	2(3)	27	27(5)	165	186(25)	0	11(7)
1.5	1	0	47	51(6)	44	51(4)	0	0(1)	59	44(15)	37	37(10)
2.5	1	0	2	4(4)	6	8(3)	12	12(4)	18	2(12)	56	43(8)
3.5	1	0	75	71(7)	51	51(6)	3	2(5)	5	2(10)	63	61(8)
4.5	1	0	73	73(7)	13	14(5)	2	4(6)	6	9(11)	13	16(7)
1.5	2	0	7	10(3)	8	11(3)	6	3(4)	0	6(9)	26	23(6)
1.5	3	0	53	49(7)	28	33(5)	34	34(7)	171	176(10)	1	9(11)
1.5	4	0	33	35(3)	19	11(3)	1	0(3)	34	45(15)	6	7(6)
1.5	-3	0	0	0(1)	0	0(1)	0	0(1)	0	0(1)	0	0(1)
1.5	-4	0	212	214(10)	188	203(11)	40	37(8)	966	1030(80)	23	45(24)
1.5	-5	0	251	263(15)	135	141(9)	8	13(7)	57	67(10)	57	61(10)
2.5	2	0	143	145(12)	86	87(5)	4	9(4)	43	43(13)	25	26(8)
3.5	2	0	28	29(3)	28	34(5)	7	3(5)	71	73(20)	38	35(9)
2.5	3	0	341	329(15)	200	189(7)	0	4(4)	89	100(23)	49	52(13)
2.5	-5	0	0	0(1)	0	0(1)	0	0(1)	--	- -	0	0(1)
2.5	-6	0	219	218(9)	92	90(6)	68	75(8)	0	2(15)	1	1(8)
-1.5	5	0	251	231(13)	135	127(7)	8	5(5)	57	38(20)	57	54(10)
4.5	-2	0	143	151(12)	86	91(6)	4	2(4)	43	40(20)	25	24(6)
-3.5	0	0	315	294(50)	187	191(18)	23	13(13)	0	0(5)	0	0(5)
-4.5	0	0	14	10(4)	13	10(3)	0	0(4)	0	0(5)	0	0(5)
-1.5	0	1	47	52(6)	40	41(4)	5	5(4)	0	4(6)	31	31(7)
-2.5	0	1	0	6(2)	1	0(1)	1	0(1)	8	6(6)	1	0(1)
-3.5	0	1	63	68(10)	81	92(7)	0	0(1)	11	10(10)	91	97(10)
-4.5	0	1	31	30(5)	22	24(4)	8	9(4)	3	0(5)	0	0(5)
-0.5	0	2	10	11(2)	5	8(3)	0	0(1)	1	3(3)	1	0(1)
-1.5	0	2	13	14(4)	18	25(4)	1	3(3)	50	41(9)	28	30(6)
-2.5	0	2	7	10(3)	6	11(3)	8	9(4)	4	7(4)	1	0(1)
-3.5	0	2	4	3(3)	3	4(3)	3	7(5)	17	9(9)	40	45(7)
-4.5	0	2	92	91(8)	13	13(3)	8	8(4)	3	2(8)	6	0(5)
-0.5	0	3	11	11(3)	15	15(4)	2	2(3)	1	3(5)	16	14(10)
-1.5	0	3	23	20(3)	11	8(4)	1	1(3)	0	4(6)	29	34(5)
-2.5	0	3	19	21(3)	3	3(3)	6	8(6)	104	94(11)	0	5(5)
-3.5	0	3	88	86(7)	69	65(5)	18	18(6)	3	12(9)	13	7(6)
-0.5	0	4	3	3(3)	1	3(3)	11	10(8)	0	2(5)	0	1(2)
-1.5	0	4	5	5(2)	12	11(2)	3	3(3)	66	85(11)	14	21(8)
-2.5	0	4	35	35(6)	15	15(3)	3	4(4)	1	11(9)	56	59(7)
-3.5	0	4	14	15(3)	-	- -	-	- -	32	30(8)	2	4(5)
-0.5	0	5	106	105(10)	63	65(6)	8	7(5)	40	47(8)	3	8(7)
-1.5	0	5	38	38(4)	24	22(6)	1	2(4)	1	7(9)	6	7(7)
-2.5	0	5	5	5(3)	16	18(3)	0	0(1)	29	30(8)	11	15(7)
-0.5	0	6	6	8(5)	30	30(5)	19	20(6)	18	20(5)	27	27(5)
-1.5	0	6	189	187(18)	90	85(8)	1	2(6)	26	26(6)	3	7(8)
R [%]			4.8		7.8		20.1		10.5		17.2	
R_w[%]			7.1		10.5		20.8		12.9		19.4	

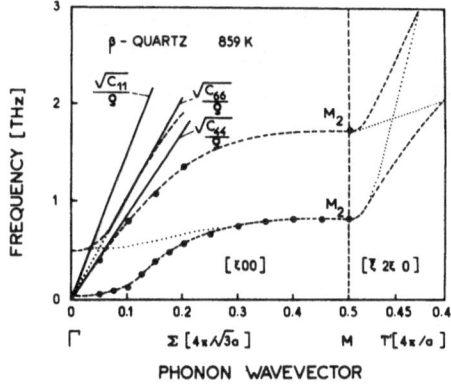

Fig. 32. The lowest phonon dispersion branches in β-quartz in Γ-M direction. Note that T' is perpendicular to Γ-M. The dotted lines indicate the supposed mode exchanges. In this interpretation the lowest branch is an optic mode. (BOYSEN et al., 1980)

(Fig. 32) is all along the dispersion curve from Γ to M a rigid tetrahedra mode which leaves the Si-O-Si bond angle constant. In the T-T' direction it is impossible that one branch always has an eigenvector of this type, because the SM going into the T direction from Γ is in the symmetric representation and the ℓβ mode going into the T' direction from M is in the antisymmetric representation. Therefore the rapid increase in frequency observed for the ℓβ mode in T' direction (Fig. 32) and expected for the soft mode in T direction is a consequence of the necessary contribution from bending the Si-O-Si bond angle.

The similarity of the ℓβ mode and the soft mode can be seen in Fig. 33. The tetrahedra 2 and 3 (around Si2 and Si3) rotate around axes in the hexagonal plane parallel

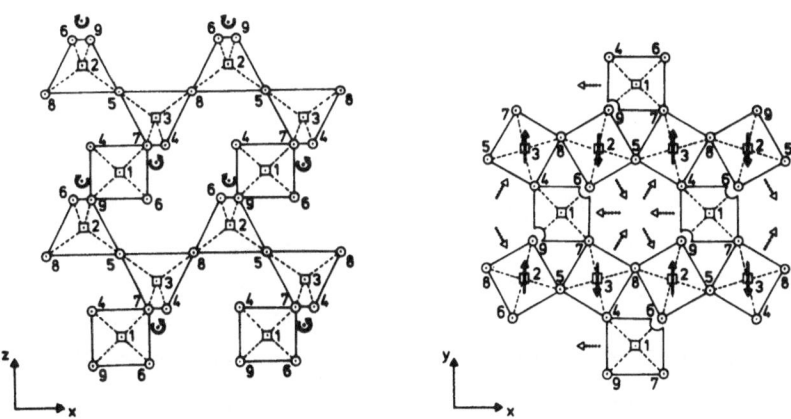

Fig. 33. The eigenvector of the lowest M-point mode in β-quartz, represented by two axes of libration (solid arrows). The corresponding translation along z of the Si-O$_4$ tetrahedron around Si(1) is not shown explicitly. The eigenvector of the Γ-point soft mode is given by its three axes of libration (dotted arrows) as well. (BOYSEN et al., 1980)

to the y axis. To obtain the movement of the soft mode, these two axes have to change
their orientation in the x-y plane such that an angle of 60° appears between the two.
Tetrahedron 1 (around Si1) has the same z component for all O atoms in $\underline{g}_1$. This is
related to the fact that the tetrahedra 2 and 3, which have the same x component and
are coupled by tetrahedron 1, have a phase of 180° between their motions. Going away
from the M point in Σ direction the wavelength in y direction increases, and thus
the phase between the mentioned tetrahedra decreases from 180°. It follows immediately that the z component of O atoms 4 and 6 drop out of phase with O atoms 7 and 9
on the coupling tetrahedron 1. This phase shift implies the onset of a rotational
component in the movement of tetrahedron 1 around an axis parallel to x, which is
just the movement in the soft mode. Therefore it is not difficult to imagine that
by change of wavelength in y direction, $\underline{g}_1$ can go over into the eigenvector of the
soft mode.

As the two investigated modes at M are always in the same representation, they
can always couple to each other. If ν_A and ν_B are the uncoupled frequencies and Δ
the coupling constant, then we obtain a dynamical matrix $\underline{\underline{D}}$ [compare (58)]

$$\underline{\underline{D}} = \begin{pmatrix} \nu_A^2 & \Delta \\ \Delta & \nu_B^2 \end{pmatrix}. \tag{65}$$

For the weakly temperature-dependent mode B we assume in simplest anharmonic approximation (LOVESEY, 1977)

$$\nu_B^2 = \nu_{B,0}^2 - b \cdot T . \tag{66}$$

For the strongly temperature-dependent mode ν_A we assume a coupling to the order
parameter δ. The essential terms of lowest order in the free energy are

$$F \sim A\delta^2 + C\delta^4 + D\delta^6 + \ldots$$
$$\nu_{A,0}^2 \cdot u_A^2 + a u_A^2 \delta^2 + \ldots \quad . \tag{67}$$

The Landau theory assumes that $A \sim (T - T_c)$ and C is negative, because the phase
transformation is of first order and appears at $T_0 > T_c$. δ is the order parameter,
i.e. the tilt angle of the SiO_4 tetrahedra (GRIMM and DORNER, 1975). u_A is the
amplitude of the uncoupled $\ell\beta$ mode at M. a is a coupling parameter of mode A to the
order parameter, and u_A^2 stands for $\sum_i u_A(q_{Mi}) u_A^*(-q_{Mi})$. This more complicated expression is necessary to keep the free energy invariant under all symmetry operations
in the β-phase. The frequency ν_A for a particular q_M we find from

$$\nu_A^2 = \frac{1}{2} \frac{\partial^2 F}{\partial u_A^2} = \nu_{A,0}^2 + a\delta^2 + \ldots \quad . \tag{68}$$

For δ^2 we used the expression (GRIMM and DORNER, 1975)

$$\delta^2 \sim 1 + \sqrt{1 - \frac{3}{4}\frac{T - T_c}{T_0 - T_c}} \quad .$$ (69)

The measured frequencies $\nu_{1,2}$ were fitted by the expression

$$\nu_{1,2}^2 = \frac{1}{2}[\nu_A^2 + \nu_B^2 \pm \sqrt{(\nu_A^2 - \nu_B^2)^2 + 4\Delta^2}]$$ (70)

where Δ was taken constant in each phase but 10 times larger in α than in β. The fitted curves are shown in Fig. 31 as full lines for the true frequencies $\nu_{1,2}$ and as dashed lines for the uncoupled frequencies ν_A, ν_B. The values $T_0 = 846$ K, $T_0 - T_c = 5$ K, $\nu_{A,0} = 0.87$ THz and $\nu_B(859$ K$) = 1.74$ THz were not varied during the course of the fit. The obtained values are $a = 1.16$ THz^2, $b = 0.0031$ THz^2/K, $\Delta(\alpha) = 1.23$ THz^2 and $\Delta(\beta) = 0.12$ THz^2. $T_0 - T_c = 5$ K gave a significantly better fit than the value 10 K which was found from the soft mode in β (AXE and SHIRANE, 1970) and from the structure (GRIMM and DORNER, 1975).

From the diagonalization of the dynamical matrix $\underline{D}$ (65) at 296 K and at 859 K it was found that the corresponding eigenvectors had rotated by about 78^0. This is in good agreement with results from the eigenvector determination, especially for the upper modes. The experimentally determined eigenvectors of the upper modes had turned by 72^0 between 573 K and 859 K and those of the lower modes by 51^0 between 296 K and 859 K.

One can summarize as follows. Two modes at the M point and in Γ-M direction exchange their eigenvectors while crossing the α-β transition regime. This exchange is not quite complete - at least between 296 K and 859 K - which, however, may not be expected. Moreover, the crystal changes its structure. A low-frequency dispersion branch in [ξ00] direction connects the strongly temperature-dependent M-point mode with the Γ-point soft mode. In the β-phase both modes maintain rigid SiO_4-tetrahedra and constant Si-O-Si bond angles. In the α-phase these two modes keep the SiO_4-tetrahedra rigid (the other M-point mode, as well).

The observations show that the Si-Si interaction, i.e. the force constant for the Si-O-Si bonding angle, is important. This agrees with the calculations of BARRON et al. (1976) who found an appreciable improvement in comparison to the older lattice dynamical models by introducing the Si-Si interaction. Barron et al. also calculated the temperature dependence of the two M-point modes concerned here by keeping the parameters of the dynamical model fixed and changing the structural parameters as they vary with temperature. A similar softening of these modes came out as observed experimentally.

5.3.4 Eigenvector Determination of the Soft Mode in $Tb_2(MoO_4)_3$

At an order-order structural phase transformation of second order the high-symmetry phase (usually high temperature) contains all symmetry elements of the low-symmetry phase (usually low temperature) and some more. In the following we will call the high-symmetry phase the paraelectric (PE) one and the low-symmetry the ferroelectric (FE) one, because the soft mode to be discussed in $Tb_2(MoO_4)_3$ is related to the PE-FE transition in this improper (expression to be explained later) ferroelectricum. The soft mode in PE generally transforms like +1 under all symmetry operations of PE which are maintained in FE, and it transforms like -1 under all symmetry opera- tions of PE which disappear at the phase transition PE → FE.

$Tb_2(MoO_4)_3$ goes from a paraelectric tetragonal ($P\bar{4}2m$) to a ferroelectric ortho- rhombic (Pba2) structure in nearly second order with doubling of the unit cell (JEITSCHKO, 1972). In the case of a doubling of the unit cell superstructure, Bragg reflexes must appear in FE at positions in reciprocal space which have been boun- daries of the Brillouin zone in PE. It is the M point (0.5, 0.5, 0) in this case. A soft mode is then expected in PE at the M point. Very generally a Brillouin zone boundary mode is "antiferrodistortive" (antiferroelectric) and cannot directly pro- duce a spontaneous polarization. The order parameter η in FE has the displacement pattern given by the eigenvector of the soft mode in PE. The actual displacements of the atoms in FE contain, besides the pattern of the order parameter, further com- ponents which come in by higher order coupling of the condensing order parameter to other modes. In the case of $Tb_2(MoO_4)_3$ the order parameter couples to the shear strain (which makes FE orthorhombic); the shear strain, in turn, produces a spon- teneous polarization by piezoelectric coupling. This material is called an improper ferroelectric because the polarization is not identical with the order parameter but proportional to η^2.

In FE this substance is also ferroelastic. This means that the orthorhombic dis- tortion can be switched from one to the other, i.e. interchanging the orthorhombic a and b axes by external stress. The spontaneous polarization which appears along the c axis changes sign upon this switching.

A soft mode was observed (Sect. 6.3) and an eigenvector determination undertaken to elucidate the mechanism of this phase transformation (DORNER et al., 1972). In PE the unit cell contains 2 formula units (34 atoms) leading to 102 branches in the phonon spectrum. In consequence, each eigenvector of these modes at a given $\underline{q}$ has 102 components. The problem was simplified by considering the MoO_4 tetrahedra as rigid, reducing the number of degrees of freedom from 102 to 48. A next step is to reduce the complexity as much as possible by symmetry considerations. Therefore a group theoretical study of the normal modes has been carried out using the multiplier- representation formalism of MARADUDIN and VOSKO (1968) together with the tables of the corresponding representations by KOVALEV (1965).

The irreducible multiplier representations of the little group of the wavevector $q_M = (0.5, 0.5, 0)$ in PE contains 8 symmetry operations R_i and 4 complex one-dimensional (non-degenerate) and one two-dimensional (doubly degenerate) representations τ_j. This fulfills the condition for irreducible representations

$$\sum_i R_i = \sum_j d_j^2 \qquad (71)$$

where d_j stands for the dimensionalities of the representations τ_j. The one-dimensional representations in the present case are pairwise conjugate complex creating the suspicion that additional degeneracies exist due to time-reversal symmetry because the eigenvectors at the M point have to be real. It turned out that the criteria for additional degeneracies (MARADUDIN and VOSKO, 1968) are fulfilled. Each pair of one-dimensional complex representations could be replaced by a real doubly degenerate representation obtained by a complex unitary transformation. These two representations are reducible. Therefore the new set of three doubly degenerate representations does not fulfill the condition (71).

The 48-dimensional eigenvector at the M point is composed of Cartesian translations and rotations (of the rigid MoO_4 groups). Projection operators have been used to decompose these eigenvectors into those which correspond to the representations. The new eigenvectors have still 48 components but only 14, 12, and 22 free parameters, respectively. The number of free parameters is identical with the number of modes per representation. As explained above, the soft mode must belong to that representation with transforms symmetrically for all symmetry operations which persist in FE and antisymmetrically for the rest. In this case the soft mode belongs to the representation which contains 12 modes. Thus 12 parameters have to be defined by an eigenvector determination.

These parameters were introduced by defining six pairs (corresponding to the two degenerate modes) of symmetry-adapted basis functions $\underline{e}_1^i$, $\underline{e}_2^i$ (Fig. 34). Note that the rotations are independent from each other only in the limit of small rotational amplitudes. The two eigenvectors $\underline{g}_1$, $\underline{g}_2$ are then

$$\underline{g}_1 = \sum_i^6 a^i \, (\underline{e}_1^i + b^i \underline{e}_2^i) \; ; \quad \underline{g}_2 = \sum_i^6 a^i \, (-b^i \underline{e}_1^i + \underline{e}_2^i) \qquad (72)$$

where the six a^i and the six b^i are the parameters to be determined experimentally.

The intensity measurements of the soft mode have been performed at $164^{\circ}C$, that is, $5^{\circ}C$ above the phase transformation. At this temperature the soft mode is heavily overdamped so that the expression F_j in (25) is reduced to a narrow Lorentzian centred at $\omega = 0$ (DORNER and COMES, 1977, p. 176). The energy transfer was sufficiently small that reliable integrated intensity measurements could be performed on a two-axis spectrometer with a 57 meV incident energy. This fact was confirmed by repeating several measurements by three-axis spectrometry. Altogether 68 "reflections" of the

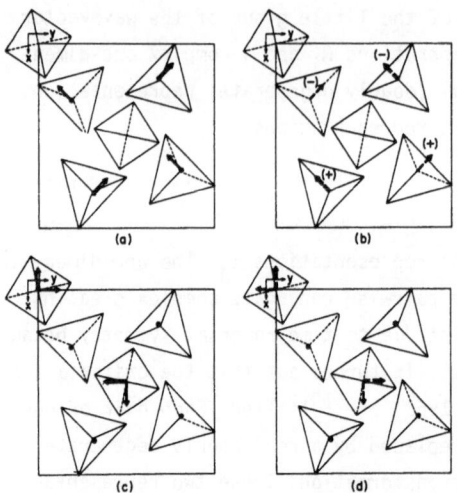

Fig. 34a-d. Basis vectors $\underline{e}^i_{1,2}$ for the eigenvector of the soft mode in $Tb_2(MoO_4)_3$ at the M point. The eigenvectors of the doubly degenerate soft mode are linear combinations of these twelve basis vectors. The double-lined arrows represent $\underline{e}^i_1$, the solid arrows $\underline{e}^i_2$. Broken arrows indicate rotations. The basis vectors are (a) $(\underline{e}^1_1, \underline{e}^1_2)$ translation of the Tb atoms and $(\underline{e}^2_1, \underline{e}^2_2)$ translation of the general MoO_4 tetrahedra; (b) $(\underline{e}^3_1, \underline{e}^3_2)$ rotation of the general MoO_4 around an axis in the x-y plane (arrows) and $(\underline{e}^4_1, \underline{e}^4_2)$ rotations about z (+ and -); (c) $(\underline{e}^5_1, \underline{e}^5_2)$ translation of the special MoO_4 tetrahedra; and (d) $(\underline{e}^6_1, \underline{e}^6_2)$ rotation of the special MoO_4. There are no translations along z. (DORNER et al., 1972)

type $(h+0.5, k+0.5, 0)$ were studied. Of these the six of type $(h+0.5, h+0.5, 0)$ should vanish for a mode in the considered representation. The small residual intensity observed at these points (Fig. 35) is ascribed to multiple scattering. Higher order contamination was negligible because a Ge(311) monochromator which does not reflect in second order was used. The remaining 62 intensities were summed pairwise, $I(h,k,0) + I(k,h,0)$ because they are equivalent in the tetragonal structure of PE. The agreement within pairs was about 5%. Thus one has finally 31 independent intensities, which could be used to fit the 12 parameters of the soft mode. This ratio of independent intensities to free parameters is not large. Therefore, one used an initial estimate for these parameters from the static displacements $\underline{U}$ as derived from the difference in atomic positions in PE and FE (JEITSCHKO, 1972).

Going from PE to FE only one of the doubly degenerate modes condenses out leading to one orthorhombic configuration in FE with the corresponding polarization. If the other mode condenses out the orthorhombic a and b axes will be interchanged and the polarization opposite. Thus the two modes are connected to the two different

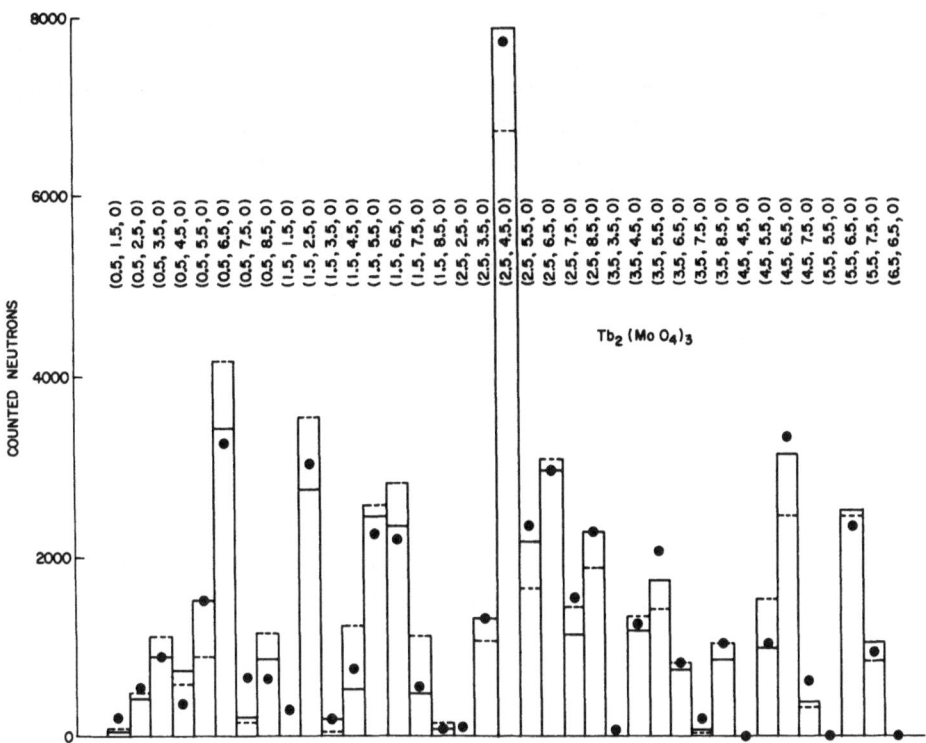

Fig. 35. Intensities of the doubly degenerate soft mode in $Tb_2(MoO_4)_3$ at the M point obtained in PE at $164^\circ C$ ($5^\circ C$ above T_0) presented by the full dots. The dashed bars (22% R factor) were calculated with soft-mode eigenvectors derived from static displacements in FE. The solid bars (9.2% R factor) are the results for a best fitting of the twelve free parameters (DORNER et al., 1972)

configurations which can be switched from one to the other by external stress or an applied electric field.

The displacements $\underline{U}$ contain 51 parameters which are the Cartesian components of the atomic displacements of one molecular unit. The positions of the other 3 molecular units in FE are defined by symmetry. Describing $\underline{U}$ by $\underline{\sigma}_1$ and using a least squares fit, a set of the 12 static parameters a^i, b^i was determined. It turned out that the eigenvector $\underline{\sigma}_{1ST}$ could not fully describe the pattern of the static displacements - a fact which had to be expected. The structure of FE was determined at room temperature ($130^\circ C$ below the phase transformation). At this temperature the coupling of the order parameter to the strain and of the strain to the polarization produce additional components in $\underline{U}$ which can not be described by $\underline{\sigma}_{1ST}$. For example, the z component of the Tb atoms and the MoO_4-tetrahedra which produce the spontaneous polarization cannot be reproduced by $\underline{\sigma}_{1ST}$. In $\underline{\sigma}_1$ these components are zero by symmetry. The higher order coupling of the order parameter to other modes in the low-symmetry phase has been extensively studied in quartz (AXE and SHIRANE, 1970; GRIMM and DORNER, 1975).

The set of the static parameters has been used as an initial guess to fit the 31 inelastic integrated intensities including the contributions from $\underline{\sigma}_1$ as well as from $\underline{\sigma}_2$. Initially the displacements were fixed at these values and the phonon reflections fitted by adjusting only a scale factor and two spherical Debye-Waller factors (one for the oxygens, and one for the remaining atoms). The result of the fit is shown in Fig. 35 as dashed bars. The agreement is already quite reasonable (22% R factor). If one then adjusts the 12 displacement parameters (due to correlation effects not all 12 could be varied simultaneously) the R factor was further reduced to 9.2% and the eigenvector $\underline{\sigma}_{1SM}$ of the soft mode was obtained. The difference between $\underline{\sigma}_{1ST}$ and $\underline{\sigma}_{1SM}$ is another strong indication that the displacements $\underline{U}$ cannot completely be described by the order parameter having the eigenvector $\underline{\sigma}_{1SM}$ of the soft mode. Contributions from other A_1 modes having the same symmetry as $\underline{\sigma}_{1SM}$ are added by the condensation of the order parameter.

This eigenvector determination of a soft mode visualises the close relation between the dynamic displacement pattern of the soft mode $\underline{\sigma}_{1SM}$ and the static displacement pattern $\underline{U}$. But it shows at the same time that the superstructure developing in the low-temperature phase is not simply proportional to the condensing order parameter.

6. Analysis of Phonon Line Shapes

As far as phonon frequencies and integrated intensities had to be determined (Chap. 5) there was no demand for high resolution. Resolution had to be sufficient in q to observe the Kohn anomalies and sufficient in ω to separate neighbouring branches. If one wants to study line shapes of phonon response functions one always asks for high resolution. Often one takes a compromise between loss in intensity and gain in resolution (Sect. 2.1). Energy resolution is generally proportional to the energy itself such that a certain ΔE is much easier obtained with small incoming energies than with large ones. This fact eases the investigations of critical phenomena (if the necessary momentum transfer is not to big). But line shapes of phonons have to be studied at the phonon frequencies which may be high.

The limited resolution of traditional inelastic neutron scattering is the reason why anharmonicity, manifesting itself in line broadening and deforming, is rarely studied.

We wish to stress that mode crossing and eigenvector exchange, as discussed in Sects 5.2 and 5.3.1, are effects which rest within the harmonic description of lattice dynamics. The eigenvectors of the diagonalised dynamical matrix are still orthogonal to each other. The introduction of a self-energy Π (24) accounts for anharmonic effects.

Microscopically, anharmonicity is created by contributions to the interatomic potential of higher than quadratic order in the atomic amplitudes. Clearly these contributions become more important at larger amplitudes, i.e. at higher temperatures. These higher order terms in the potential create a coupling and energy transfer between different phonons.

The "normal" anharmonic effect of frequency shift and line broadening is examined on AgBr (Sect. 6.1). The pathological anharmonic case of a soft mode is presented in Sect. 6.2. A very spectacular case of a double-peak structure due to anharmonicity at 4.2 K was studied in CuCl (Sect. 6.3).

6.1 Frequency Shift and Damping in AgBr

Silver halides show thermally activated Frenkel disorder in the cation (Ag^+) sub-lattice. At very high temperatures they exhibit an anomalously high ionic conductivity. At temperatures 100 to 150 degrees below the melting point it rises above the values extrapolated from the low-temperature regime. A 1 cm^3 cube of AgBr has a resistance of only a few ohms at the melting point. This behaviour is attributed to either long-range Coulomb interaction between the defects or to a general softening of the crystal lattice.

Recent ionic conductivity measurements that have been carefully analysed (ABOAGYE and FRIAUF, 1975) suggest that long-range Coulomb interaction is not sufficient to account entirely for the observed anomaly. Rather, the defect concentration in AgBr of the order of 1% close to the melting point seemed to indicate that the lattice arrangement of atoms is strongly disturbed at high temperatures. A possible mechanism for the migration of the Ag^+ ions that could explain the conductivity enhancement is the formation of dumb-bell interstitials of two silver ions oriented along the <ξξξ> direction between two Br layers (KLEPPMANN, 1976; KLEPPMANN and BILZ, 1976). The quadrupolar deformability of the silver electronic shell that was recently introduced to account for several anomalous properties in the phonon spectra (FISCHER et al., 1972; Von der OSTEN and DORNER, 1975; DORNER et al., 1976) supports this model.

To provide more experimental data necessary for a better understanding of the premelting phenomena, DORNER et al. (1977) studied certain lattice phonons in AgBr at high temperatures by using inelastic neutron scattering. In particular, we have measured some acoustic branches along different directions in the Brillouin zone paying attention to the change in frequency and broadening of the phonon peaks with temperature. Since the TA branches should be sensitive to the silver ion interstitials, particular effects were expected for these. We searched carefully for a double-peak line shape which would be related to the interaction of the lattice phonons with a local mode to be expected, e.g. from the silver dumb-bell configuration. The sample was kept in vacuum and heated by thermal radiation. Spectra were taken at 293, 523, 573, 623 and 673 K, the melting point of AgBr being at 698 K.

While we rapidly lost the signal for optic modes, we concentrated on measuring essentially the transverse acoustic branches along <ξ00>, <ξξ0>, and <ξξξ> at several temperatures. In Figs. 36 and 37 representative phonon scans are pictured. The full line in each scan is obtained by fitting the phonon peak by a damped harmonic oscillator function (25) and the elastic peak at $\omega = 0$ by a Gaussian. The eigenfrequencies of the harmonic oscillator in each case are indicated by the arrows.

As seen in the figures, the temperature behaviour of the two TA branches are found to be quite different. In the <ξ00> direction (Fig. 36) the phonon frequencies decrease only slightly with temperature and do not broaden significantly. In con-

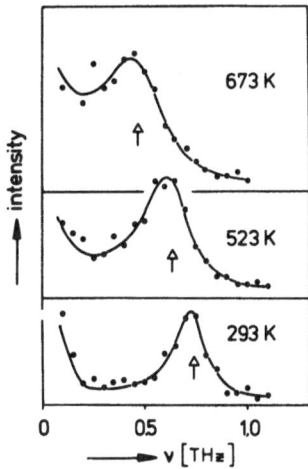

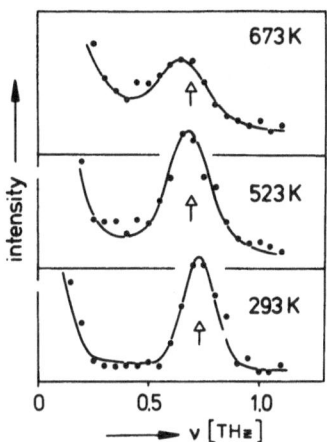

Fig. 36. Constant-Q scans in AgBr in
[ξ00] direction with ξ = 0.6 for the
TA phonon at different temperatures.
The full curves are the result of a
fit of (25) plus Gaussian around ν = 0.
The arrows indicate the quasi-harmonic
frequency ν_j [ω_j in (25)]. (DORNER et
al., 1977)

Fig. 37. Constant-Q scans in AgBr in
[ξξξ] direction with ξ = 0.2 for the
TA phonon at different temperatures.
See the explanation in Fig. 36

trast, the <ξξξ> phonon (Fig. 37) exhibits a stronger change in frequency and broadening. The comparison for different q-values within one phonon branch suggests the
relative frequency change with temperature to be constant. Giving the frequencies
in percent of the frequencies obtained in previous measurements at liquid nitrogen
temperature (DORNER et al., 1976) the TA <ξξξ> and the TA <ξ00> decrease to about
63% and only 90%, respectively, at 673 K. Similarly, the <ξξ0> phonons were found
to decrease to 56% (for polarization along [1$\bar{1}$0]) and 90% (for polarization along
[001]) for the same temperature. Since the decrease has the same percentage for all
phonons along a particular branch it is possible to describe the temperature dependence of the frequency by only one sound velocity per phonon branch and temperature.
The well-known relations between the sound velocities and the elastic constants
allow computing c_{11}, c_{12}, and c_{44} at the corresponding temperatures. In Fig. 38 our
results are represented together with earlier low-temperature data (MARKLUND et al.,
1977) obtained by an ultrasonic technique. Both sets of data show reasonable agreement within the limits of error, that is, are different due to the various experimental techniques used. While c_{44} is small and nearly independent of temperature,
leading to the well-known violation of Cauchy relation, c_{11} and c_{12} show a significant decrease with increasing temperature.

The phonon damping as obtained from our fit increases at higher temperatures.
Fig. 39 shows that the damping constant Γ within one single phonon branch is pro-

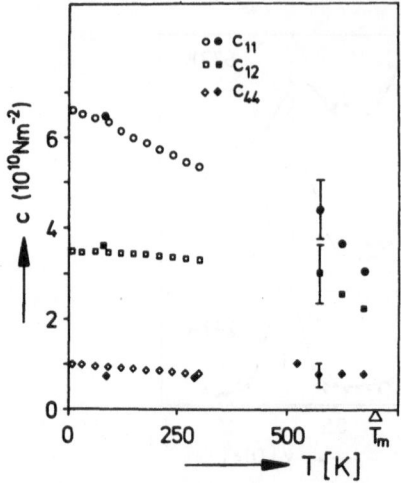

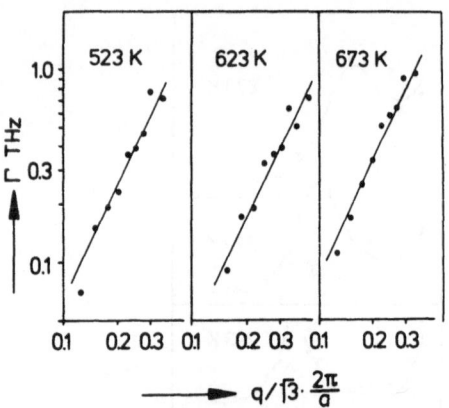

Fig. 38. The elastic constants of AgBr vs. temperature. Open symbols represent data obtained by MARKLUND et al. (1977) and full symbols represent results from inelastic neutron scattering by DORNER et al. (1977)

Fig. 39. The damping constant Γ for the TA phonon in AgBr in [$\xi\xi\xi$] direction. was obtained at different phonon wavevectors q and at different temperatures by fitting (25) to the experimental data. Γ is plotted in a doubly logarithmic scale versus q for different temperatures. The solid lines correspond to $\Gamma \sim q^2$ (DORNER et al., 1977)

portional to q^2. At constant q, in the temperature interval 293 K to 673 K the damping increases by about a factor of 2, the statistical error, however, being too big to allow for giving quantitative results. In particular, we were not successful in observing any indication for a double hump structure of the phonon scans outside of statistics.

As shown by KLEPPMANN (1976) the anomalously high ionic conductivity in AgBr results essentially from the small value of the elastic shear constant c_{44}, which in turn is a direct consequence of the quadrupolar deformability of the silver ion (FISCHER et al., 1972). Compared to corresponding alkali halides this generally leads to a low value of the activation energy for vacancy and interstitial diffusion in agreement with experimental observation. The model developed by KLEPPMANN and BILZ (1976) explicitly shows the influence of the elastic constants onto the activation energy given by

$$E \sim (c_{11} - c_{12}) + 3c_{44} \; . \tag{73}$$

For increasing temperatures our measurements (Fig. 38) suggest a further decrease of ($c_{11} - c_{12}$). This gives rise to an additional lowering of the activation energy at elevated temperatures and is qualitatively able to explain the increase in conductivity. From the analysis of our data a 27% decrease in activation energy follows from 293 K to 673 K.

Discussing the phonon attenuation the observed q^2 dependence (Fig. 39) is well understood within the framework of current theories describing anharmonic interactions (KWOK, 1967; WOODRUFF and EHRENREICH, 1961). At elevated temperatures we are apparently in the first sound regime $\omega\tau \ll 1$, where ω is the frequency of a particular phonon and τ the relaxation time of the local phonon density. τ decreases with increasing temperature. A particular phonon with frequency ω may be in the first sound regime at high temperature and in the zero sound regime $\omega\tau \gg 1$ at low temperature. In the first sound regime the local temperature given by the local phonon density oscillates adiabatically with the phonon frequency ω. In the zero sound regime the local temperature is the same as the macroscopic temperature. For the zero sound regime theory predicts that Γ is proportional to q. Usually the sound velocities of the two regimes are different. This has been studied extensively by LOIDL et al. (1976a, 1976b) in NaF.

6.2 Structural Lattice Instabilities

The most spectacular anharmonic effect represents the soft mode observed in connection with structural order-order phase transformation. As explained in Sect. 5.3.4 the soft mode in the high-symmetry phase has an eigenvector which is the same as the eigenvector of the static order parameter condensing out in the low-symmetry phase. It is plausible that the restoring forces for this mode will decrease on approaching the phase transformation and consequently the frequency of this mode will decrease and get soft. This soft-mode frequency approaches zero for a second-order phase transformation as $T \to T_c$ (Sect. 6.2.1). Sometimes it levels off at a finite frequency and a "central peak" at $\omega = 0$ appears (Sect. 6.2.2). Historically the concept of a soft mode was introduced from the temperature dependence of the dielectric constant at para-ferroelectric phase transformations by COCHRAN (1959a) and independently by ANDERSON (1960).

For the investigation of these soft modes in the high-symmetry phase the inelastic scattering of neutrons is still a unique technique. For ferrodistortive transformations the soft mode (at the centre of the Brillouin zone) is very often Raman and infrared inactive. For antiferrodistortive transformations the soft mode appears at a boundary of the Brillouin zone, and is therefore accessible by inelastic neutron scattering only.

Soft modes have been observed in many substances and described by (25). As long as the mode is not overdamped ($2\omega_j^2 > \Gamma_j^2$) the scattered intensity exhibits two maxima, one for energy gain and one for energy loss. In the overdamped case ($2\omega_j^2 < \Gamma_j^2$) there is only one maximum at $\omega = 0$. The transition from underdamped to overdamped soft modes at structural phase transformations is always found to be caused by the decrease of ω_j. Within experimental error, Γ_j has always been found to be temperature independent.

6.2.1 Soft Mode in $Tb_2(MoO_4)_3$

The soft mode investigated in $Tb_2(MoO_4)_3$ ($T_0 = 159^oC$) (DORNER et al., 1972) at the
M point (Fig. 40) at high temperatures could be described by (25) (Fig. 7). Below
400^oC the mode becomes overdamped $[\Gamma_j^2 > 2\omega_j^2(q)]$, as seen in Fig. 41. At temperatures
above 400^oC, ω_j was determined by a least-squares fit to the shape of the response,
fitting Γ and ω_j simultaneously (Fig. 42). Below 400^oC this was no longer possible.
The data near T_0 were obtained from the integrated intensity (27) which holds for
the damped and overdamped harmonic oscillator as well.

Apparently the Curie-Weiss law is well fulfilled. We might call $1/\omega_j^2$ the structural susceptibility of the system corresponding to the order parameter η. $Tb_2(MoO_4)_3$
is one of the rare examples where the soft mode moves out of the overdamped regime
with increasing temperature and becomes easily resolvable at high temperatures.

The Curie-Weiss law, which was first derived for the magnetic susceptibility of
a ferromagnetic material in the paramagnetic phase, apparently describes the temperature dependence of the soft-mode frequency in $Tb_2(MoO_4)_3$ very well. This is not
always the case, as we will show later. To understand these facts a bit better we
consider the potential energy V of the lattice as a function of the amplitudes A_j
of the normal modes j. Expanding in the amplitudes A_j up to fourth order we get

$$V = \sum_j \frac{1}{2} \omega_{0,j}^2 A_j^2 + \sum V_{j_1 j_2 j_3 j_4}^{(4)} A_{j_1} A_{j_2} A_{j_3} A_{j_4} + \dots \quad . \tag{74}$$

The third-order term, treated properly (LOVESEY, 1977), has a temperature dependence
as the fourth-order term and is not written explicitly here. This complicated expression is simplified by taking the mean potential experienced by the jth normal
mode $<V_j>$.

$$<V_j> = \frac{1}{2} \omega_j^2 A_j^2$$

$$= \frac{1}{2} \omega_{0,j}^2 A_j^2 + \sum_{j_1} V_{j_1 j_1 jj}^{(4)} <A_{j_1}^2> A_j^2 + \sum_{j_1} V_{j_1 j_1 jj}^{(4)} (A_{j_1}^2 - <A_{j_1}^2>) A_j^2 + \dots \quad . \tag{75}$$

The cross products between normal coordinates of different modes drop out since
normal modes are independent, at least in lowest order. The last term in (75) ex-
presses the fluctuations in frequency which introduce a damping Γ_j.

The thermal average of the amplitudes $<A_{j_1}^2>$ is assumed to be proportional to the
Bose occupation factor $[n(\omega_{j_1}) + 1/2]$. This identity is strictly true only in the
harmonic case. A better calculation has to be self-consistent, meaning that the fluctuations and the temperature dependence of ω_{j_1} (anharmonic contributions to the
other modes) should be included. This is partly done in calculations by MIGONI et
al. (1976, 1978) which will be discussed soon.

For non-pathological modes, $\omega_{0,j}^2$ is positive and quite large, so that the anharmonic contribution of the second term in (75) is relatively small. This contribution

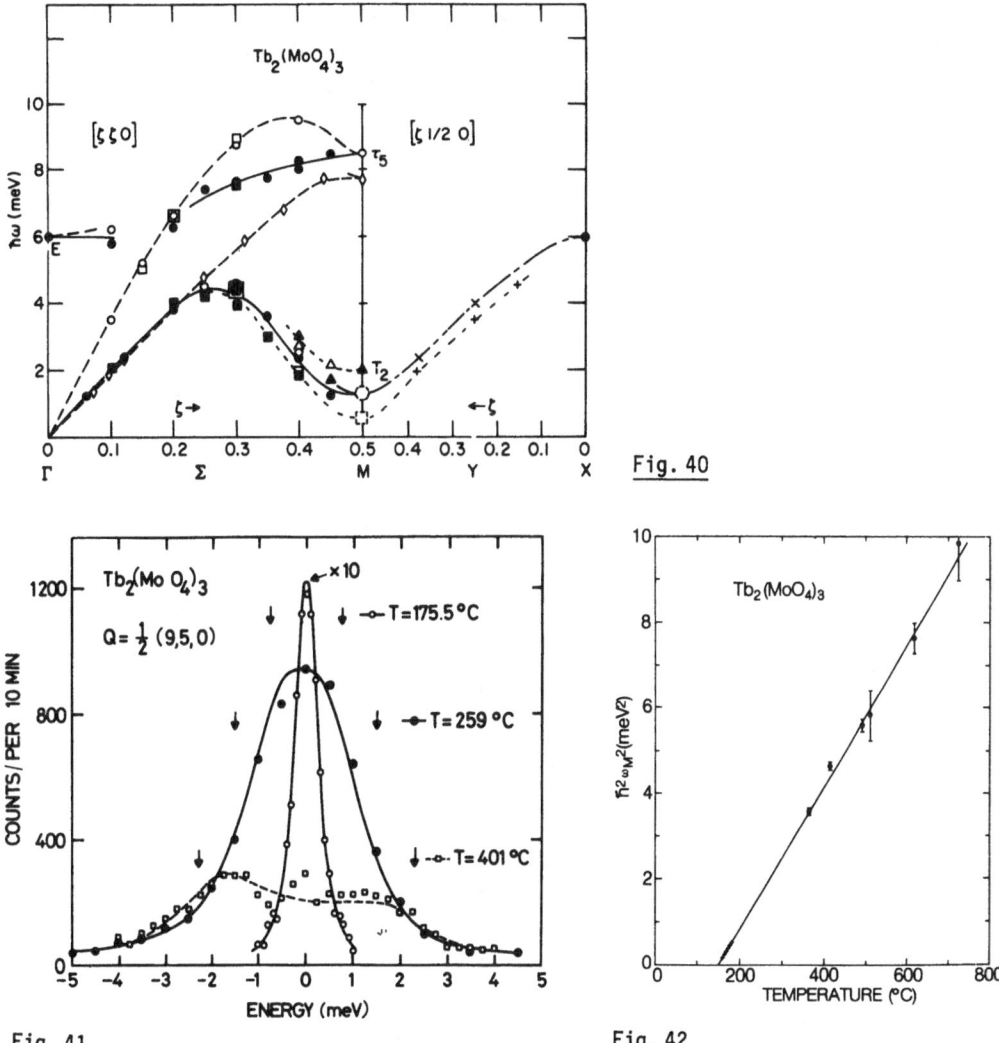

Fig. 40

Fig. 41

Fig. 42

Fig. 40. Phonon dispersion curves in $Tb_2(MoO_4)_3$ near the M point. Different symbols indicate different scattering geometries, different sample orientations, and different temperatures. The temperature dependence of the soft mode at the M point is clearly seen. At 414°C (△) the soft mode is underdamped (compare Fig. 7); at 260°C (o,●,×) and at 184°C (□,+) the soft mode is overdamped at the M point but not for symbols (●,×,+) (DORNER et al., 1972)

Fig. 41. Constant-Q scan through the soft mode in $Tb_2(MoO_4)_3$ at the M point at different temperatures. The solid lines represent a fit of (25) modified with the resolution normalization factor (16). The arrows indicate the harmonic frequencies ω_M [ω_j in (25)]. Note the critical behaviour of the diverging intensity at $\omega = 0$ (DORNER et al., 1972)

Fig.42. Temperature dependence of the frequency ω_M of the soft mode in $Tb_2(MoO_4)_3$ at the M point. At higher temperatures ω_M was determined by fitting the spectral profiles, (Fig. 7) to (25). The lower temperature data (×) were obtained from the temperature dependence of the integrated intensity (27). The straight line represents the best fit of the Curie-Weiss law $(\hbar\omega_M)^2 \sim (T - T_c)$. (DORNER et al., 1972)

is usually negative. But it might be that $\omega_{0,j}^2$ is negative and ω_j^2 is only positive because the second term in (75) is large and positive. The unstable harmonic mode $\omega_{0,j}$ is stabilized due to the anharmonic coupling to the averaged amplitudes of all the other modes.

The temperature dependence of ω_j comes from those modes for which $\hbar\omega_{j_1} \ll k_B T$; then $\langle A_j^2 \rangle$ is proportional to T. At T_c where ω_j becomes zero the second term is equal to the first in (75).

The Curie-Weiss law is only obtained in substances which have a sufficiently high T_c, such that enough modes ω_{j_1} are thermally occupied. At low temperatures the second term in (75) is not linear in temperature anymore, because the $\langle A_j^2 \rangle$ approaches the zero-point amplitudes. Therefore the anharmonic stabilizing contribution approaches a constant value at low temperatures. The behaviour of the ferroelectric soft modes in $KTaO_3$ and in $SrTiO_3$ which never condense out are explained in this way by MIGONI et al. (1976, 1978). ω_j^2 is in these cases linear in temperature only at high temperatures with a slope which extrapolates to a positive T_c (Fig. 43).

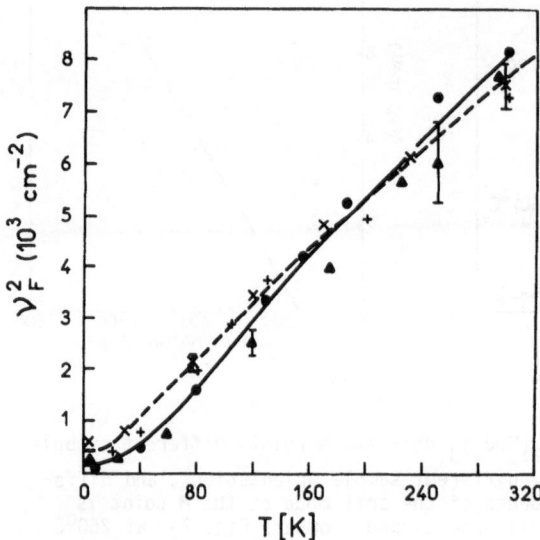

Fig. 43. Temperature dependence of the ferroelectric soft mode ν_F in $KTaO_3$ ($\times$, +, ---) and $SrTiO_3$ ($\bullet$, $\blacktriangle$, —). The drawn curves are calculated on the basis of (75) but include a microscopic model by MIGONI et al. (1976, 1978)

There are only a few cases where the microscopic origin of the unstable mode $\omega_{0,j}$ is discussed with the exception of low-dimensional metals (COMES and SHIRANE, 1979), where the energy of the conduction electrons stabilizes the lattice. Similar to this in the sense of electron lattice interaction is the cooperative Jahn-Teller effect as discussed by BIRGENEAU et al. (1974) for the 151 K phase transformation in $PrAlO_3$. Recently MIGONI et al. (1976, 1978) gave a microscopic explanation for the ferroelectric soft mode in the insulators $KTaO_3$ and in $SrTiO_3$ by introducing an anisotropic polarisability of the oxygen ion.

6.2.2 The Central-Peak Phenomenon

The "central peak" was discovered by RISTE et al. (1971) near the displacive phase transformation in $SrTiO_3$. At the time the phenomenon of a "soft mode" was well-known. The striking feature of the central peak is that an intensity appears at $\omega = 0$ while the two soft-mode maxima are very well separated. Recently an extended review on central-peak phenomenon has been published by MUELLER (1979) discussing various experimental techniques as well as different theoretical approaches.

The existence of the central peak in $SrTiO_3$ at $(1/2, 1/2, 3/2)$ was proved by RISTE et al. (1971) because they used the time-of-flight technique which separates the higher order contributions $(2k_I, 3k_I, ...)$. Other groups had overlooked the effect because it was hidden under the intensity coming from $2k_I$ being Bragg scattered from (113). Afterwards, the use of monochromators and analysers which do not reflect in second order like Ge (111), and the use of good pyrolytic graphite filters allowed the investigation of the central peak with Three-Axis-Spectrometers (SHAPIRO et al., 1972).

The experimental results show that at low frequencies there is a response in addition to that which is described by the "soft-mode" formula (25) where the damping is proportional to ω. To describe this additional response the self-energy expression can be extended by assuming that the soft mode decays into some other mode or combinations of modes and that this further mode decays exponentially with a long characteristic time $1/\gamma$ (COWLEY, 1974). The extended self-energy [compare (24)] can be written as

$$\Pi_j(\omega,T) = \Delta_j(T) - i\omega\Gamma_j(T) - i\omega \frac{\partial^2(T)}{\gamma - i\omega} \tag{76}$$

where $\Delta_j(T)$ and $\Gamma_j(T)$ are the usual renormalization of the frequency and the usual damping constant accounting for ordinary phonon-phonon scattering. The third term describes the coupling (proportional to the coupling constant δ^2) to the mode (still not specified) introduced above. The resulting response function F_j of the scattering function reads

$$F_j(\omega,T) = \frac{\omega}{1 - \exp(-\hbar\omega/kT)} \cdot$$

$$\frac{\Gamma_j(T) + \gamma\delta^2(T)/(\omega^2+\gamma^2)}{\left[\omega_j^2(T) + \omega^2\delta^2(T)/(\omega^2+\gamma^2) - \omega^2\right]^2 + \omega^2\left[\Gamma_j(T) + \gamma\delta^2(T)/(\omega^2+\gamma^2)\right]^2} \cdot \tag{77}$$

The static structural susceptibility χ is proportional to ω_j^{-2}. For historical reasons of the central-peak phenomenon we use the following symbols

$$\omega_0^2 = \chi^{-1} = \omega_j^2 = \Omega_j^2 + \Delta_j^2 \; ; \quad \omega_\infty^2 = \omega_0^2 + \delta^2 \; ; \quad \gamma' = \frac{\omega_0^2}{\omega_0^2}\gamma \; . \tag{78}$$

Here ω_∞ is the renormalized soft-mode frequency as observed experimentally and γ' the temperature-dependent width of the central peak (due to the temperature dependence of ω_0)

$$F_j(\omega,T) = \frac{\omega}{1 - \exp(-\hbar\omega/kT)} \; \frac{\delta^2(T)}{\omega_0^2(T)} \frac{\gamma'(T)}{\omega_\infty^2(T)} \frac{\gamma'(T)}{\omega^2 + \gamma'^2(T)} + \frac{\Gamma_j(T)}{[\omega_\infty^2(T) - \omega^2]^2 + \omega^2\Gamma_j^2(T)} \tag{79}$$

which could be derived under the assumptions that

$$\Gamma_j \ll \delta^2/\gamma \; , \quad \omega_\infty^2 \gg \gamma^2 \; . \tag{80}$$

Equation (79) clearly distinguishes two contributions, one Lorentzian with width γ' around $\omega = 0$ and a well-known damped harmonic oscillator (soft-mode) expression as (25). There are two relaxation times, one proportional to $1/\Gamma_j$ and one proportional to $1/\gamma'$.

Note that SHAPIRO et al. (1972) start with a different coupling in the self-energy, but that they arrive at the same expression as (79). The only difference is that it follows from their Ansatz on the self-energy that $\Omega_j^2 + \Delta_j = \omega_\infty^2$.

The width γ' in SrTiO$_3$ has not yet been determined by inelastic neutron scattering. Therefore the question is open to what extent the central peak as observed in neutron scattering is dynamic in origin. As in SrTiO$_3$ above T_c, precursers of the new structure (soft mode and central peak) appear at the boundary of the Brillouin zone at the R point, inelastic neutron scattering is the only technique to study these phenomena. From high-resolution measurements (TOEPLER et al., 1977) it was found that $\gamma' < 20$ MHz. The current interpretation of the central peak in SrTiO$_3$ (and several other substances) is that it arises from a coupling of the soft mode to lattice defects (HALPERIN and VARMA, 1976).

Both ω_∞ and ω_0 are strongly temperature dependent. For large $T - T_c$ ω_∞^2 decreases linearly with temperature, but levels off near T_c and remains finite $\omega_\infty = \delta$ at T_c.

An interesting case in this context is KCN (ROWE et al., 1978), where γ is rather large and its origin is the reorientation of (CN)$^-$ ions. A theory by MICHEL and NAUDTS (1978) describing the coupling between reorientations and translations was used to analyse the line shapes. They arrive at the same expression as (77).

The phonon (transverse acoustic in KCN) will appear near to $\omega_j^2 + \delta^2\omega^2/(\omega^2 + \gamma^2)$. If $\omega_j^2 \gg \gamma^2$, the quasi-harmonic phonon frequency is $\omega_\infty^2 = \omega_j^2 + \delta^2$ and if $\Gamma_j \ll \delta^2/\omega_j$, then (79) describes the whole response including the central peak.

If $\omega_j^2 \ll \gamma^2$, the quasi-harmonic frequency is $\omega_0^2 = \omega_j^2$ and the central peak is absent (HALPERIN and VARMA, 1976). In this case, where ω_0 is a measurable frequency, ω_0 is the phonon frequency including the coupling between rotational reorientations and translations. In the first case ω_∞ is the phonon frequency without this coupling.

For KCN ROWE et al. (1978) found that γ was nearly independent of temperature, and Γ_j negligibly small. But in contrast to $SrTiO_3$, for KCN γ was large ($\hbar\gamma = 3$ meV), that is, in the order of acoustical phonon frequencies. The case $\omega_j^2 \ll \gamma^2$ was verified for low-frequency acoustic phonons (small phonon wave vectors q) with $\omega_j \approx 0.25\ \gamma$. As expected no central peak was observed. The case $\omega_j^2 \gg \gamma^2$ could not be verified because the phonon wavevectors corresponding to such ω_j's had to be bigger than 0.5 of the Brillouin zone, and there the coupling δ^2 goes to zero. But nevertheless the central peak was already visible for $\omega_j \approx \gamma$ at q = 0.5. For a review on inelastic neutron scattering from the central peak, see DORNER (1981a).

To describe the central peak one had to introduce a particular frequency dependence of damping (76) to produce a third peak in the response function. Normal damping is proportional to frequency (24). In Sect. 6.3 we will present a case with another particular frequency-dependent damping.

6.3 Frequency-Dependent Damping in CuCl at 5 K

CuCl has a greater ionic contribution to bonding than covalent. But it crystallises in the zinc blende structure, which is typical for covalent bonding and in which each Cu is tetrahedrally surrounded by Cl ions and vice versa. As found by PHILLIPS (1970) and Van VECHTEN (1969) CuCl is the most ionic substance having this structure, with a fractional ionic character f = 0.74, while the stability limit is 0.78. In other words, the free energy of CuCl in the rock-salt structure which is typical for ionic bonding and in which one ion is octahedrally surrounded by other species should be very close to the one it has in zinc blende. As those two structures may be related by a displacive transformation in which, for example, the Cl ions move from their (1/4 , 1/4 , 1/4) positions to (1/2 , 1/2 , 1/2), the vibrational potential is expected to be strongly anharmonic having large contributions, higher than quadratic in the ionic amplitudes even for small amplitudes at low temperature. The anharmonic part in the potential alone does not produce dramatic anharmonic effects by itself. Only if a mode is connected to a particular distribution of decay channels (combinations of other phonons which couple to the first one with conservation of momentum and energy) the phonon response may not only be broadened but more drastically deformed.

By light scattering from CuCl at low temperatures [for references see (HENNION et al., 1979)] three peaks were observed in the frequency spectrum instead of the expected two, one TO and one LO. Polarised Raman experiments by KRAUZMANN et al. (1974) at 40 K have shown that two of the peaks have essentially a TO character, so that it is impossible to assign, for instance, one of the peaks to the TO zone-centre phonon and the second to some second-order Raman scattering. To explain this result, a model was built in which a very strong third-order interaction between the TO phonon with frequency ω_{TO} and a two-phonon density of states with a postulated singularity at ω_c near to ω_{TO} was assumed.

Phonon dispersion curves of CuCl at 4.2 K have been measured by inelastic neutron scattering and fitted to a shell model by PREVOT et al. (1977) (Sect. 4.1). The two-phonon density of states for $g_1 - g_2 \approx 0$ was calculated from the fitted shell model (Fig. 44). Apparently this density of states has a sharp cut-off at $\omega_c = 5.11 \pm 0.02$ THz which is the value postulated to interpret the polarised Raman experiment by KRAUTZMANN et al. (1974).

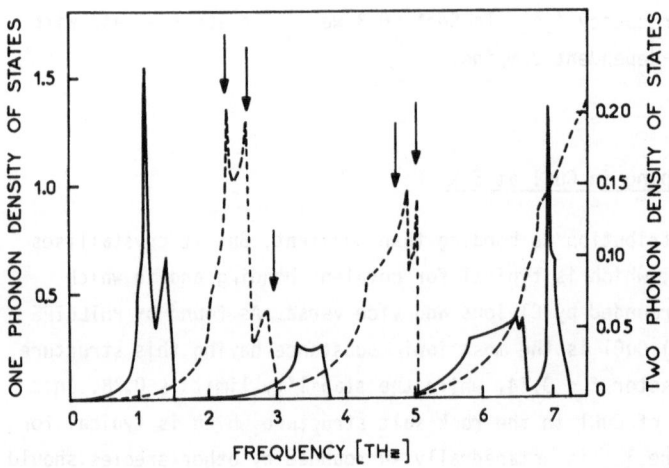

Fig. 44. Phonon density of states in CuCl at 4.2 K calculated on the basis of a shell model fitted to the experimental data (Fig. 11). Full lines represent one-phonon density of states and broken lines a two-phonon density of states $\gamma(\omega)$ calculated for combinations of phonon wavevectors $g_1 = -g_2$ (PREVOT et al., 1977). Thus this two-phonon spectrum can couple to modes in the zone centre at $g = 0$. The arrows indicate main features in the infra-red absorption spectrum of CuCl at 2 K (IKASEWA, 1973)

Further measurements with inelastic neutron scattering which is insensitive to second-order scattering have been performed to prove that the double-peak structure of the TO phonon is a first-order effect at $q = 0$ and to follow the effect into the Brillouin zone for $q \neq 0$, where the frequency difference $\omega_{TO}(q) - \omega_c(q)$ varies (HENNION et al., 1979).

The required resolution to separate the two maxima, one broad at 4.53 THz and one narrow at 5.13 THz, and the absorption of Cl (33.6 barns) rendered the experiment time-consuming. The results at $q = 0$ (at 3,3,3) are presented in Fig. 45, where

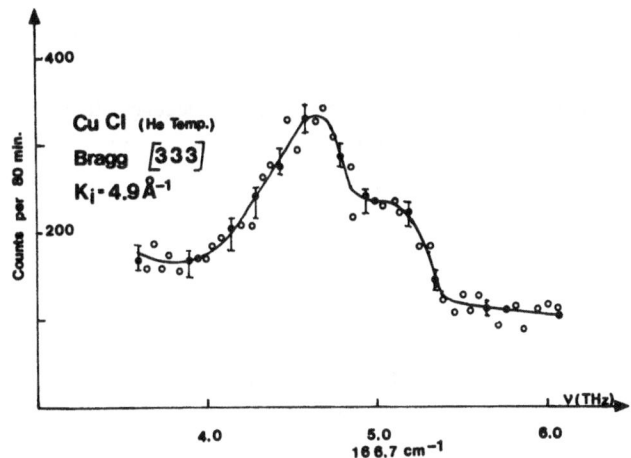

Fig. 45. Constant-Q scan through the TO mode at q = 0 in CuCl at 4.2 K. The curve is a guide line to the eye (HENNION et al., 1979)

a double structure at the right frequencies is clearly visible. To prove that this structure is not a spurious effect, we performed similar scans at different neutron energies and at different points in reciprocal space. All these scans gave the same profile and intensities as calculated from the dynamical structure factor (19). For the final analysis at q = 0 all results have been averaged. The relatively high background came predominantly from the sample and was attributed to incoherent scattering.

To describe this response theoretically, we recall that it is proportional to the imaginary part of the dynamic phonon suceptibility χ_{Ph} (21). In case the absolute contributions from the self-energy (24) are small compared to the harmonic frequency Ω_j, as given here, we can simplify (21) to

$$\chi_{Ph}[\Omega_j(\underline{q}),T,\omega] = \{2\Omega_j(\underline{q}) \, [\, \Omega_j(\underline{q}) - \omega + \Pi_j(\underline{q},T,\omega)/2\Omega_j(\underline{q})]\}^{-1} \ . \tag{81}$$

Then (25) is replaced by

$$F_j[\omega,\Omega_j(\underline{q}),T] = \frac{1}{1 - \exp(-\hbar\omega/kT)} \, \frac{1}{2\Omega_j(\underline{q})} \, \frac{\Gamma_j(\underline{q},\omega,T)}{[\Omega_j(\underline{q}) + \Delta_j'(\underline{q},\omega,T) - \omega]^2 + \Gamma_j^2(\underline{q},\omega,T)} \ . \tag{82}$$

Here Ω_j has to be taken positive for phonon creation and negative for phonon annihilation, and Δ_j' is defined by

$$\Delta_j'(\underline{q},T) = \Delta_j(\underline{q},T)/\Omega_j(\underline{q}) \ . \tag{83}$$

At a low temperature $\hbar\omega \gg kT$, F_j is proportional to

$$F[\omega,\Omega(\underline{q})] \sim \frac{\Gamma(\underline{q},\omega)}{[\Omega(\underline{q}) + \Delta'(\underline{q},\omega) - \omega]^2 + \Gamma^2(\underline{q},\omega)} \frac{1}{\Omega(\underline{q})} \,. \tag{84}$$

The suffix j identical to TO is dropped. Now one has to define the frequency-dependent quantities Δ' and Γ which are related by the Kramers-Kronig transformation.

One starts to evaluate Γ from the two-phonon density of states $\gamma(\omega)$ at q = 0 (Fig. 44), approximated by the elliptic form

$$\gamma(\omega) \sim D \{1 - [(\omega - \omega_c - a)/a]^2\}^{1/2} \tag{85}$$

which has a sharp cut-off at ω_c. A kind of spectrum of decay channels was obtained by assuming an interaction inside the two phonon continuum by fourth-order anharmonic interaction which enhances the high-frequency part of $\gamma(\omega)$. The damping $\Gamma(\omega)$ at q = 0 was then obtained by multiplying this spectrum of decay channels by a third-order anharmonic coupling parameter of the TO mode. This parameter was assumed constant over the considered frequency range ω_c to $\omega_c - 2a$. In Fig. 46 $\Gamma(\omega)$ is presented together with $\Delta'(\omega)$, which was derived by the Kramers-Kronig transformation (KRAUZMAN et al., 1974). The response function F (84) will have maxima at

$$\Omega + \Delta'(\omega) - \omega = 0 \tag{86}$$

or in the vicinity. Therefore the straight line $\omega - \Omega$ is drawn in Fig. 46a. Intersections of this line with $\Delta'(\omega)$ provide ω values which fulfill (86). There are three such values, ω_0 , ω_1 , ω_2 .

At ω_0 the damping is zero ($\omega_0 > \omega_c$) leading to a δ function. At ω_1 the damping is very strong, so that no visible maximum appears. A bit below ω_2 where $\Gamma(\omega)$ is considerably reduced a maximum appears. As seen from Fig. 46b this calculation describes the Raman data very well (KRAUZMANN et al., 1974).

An intriguing question is: what will happen if the value of the harmonic frequency Ω could be shifted without changing $\Delta'(\omega)$ and $\Gamma(\omega)$?

A Raman experiement under hydrostatic pressure (SHAND et al., 1976) revealed that with increasing pressure the sharp line moved to higher frequencies and got more intensity at the expense of the broad maximum. This was explained by the above theory assuming that Ω increases faster with increasing pressure than ω_c of the two-phonon spectrum.

In inelastic neutron scattering we have the opportunity to perform scans at $\underline{q} \neq 0$. As can be seen from Fig. 11 the frequency of the TO branch increases with $\underline{q}$ for all three symmetry directions. The two-phonon spectrum for small $\underline{q} = \underline{q}_1 + \underline{q}_2$, where $\underline{q}$ is the wavevector of the TO mode and $\underline{q}_1$ and $\underline{q}_2$ the wavevectors of the two interacting acoustic phonons, will be nearly constant because the acoustic dispersion curves are very flat within some distance from the Brillouin zone boundary. If any change in $\omega_c(\underline{q})$ is expected, it should decrease with increasing $\underline{q}$.

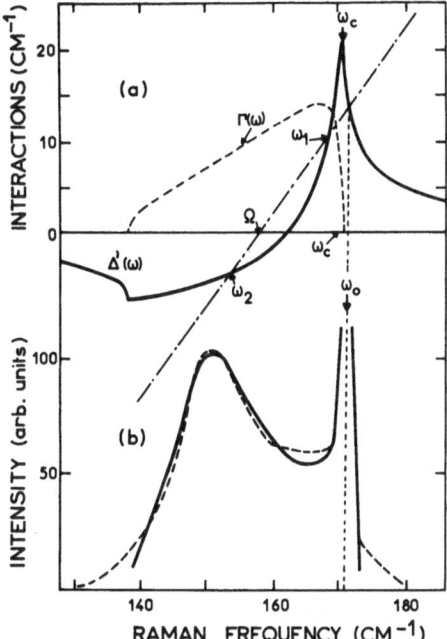

Fig. 46a,b. The anharmonic coupling of the TO mode in CuCl. (a) The dashed curve gives the frequency-dependent damping $\Gamma(\omega)$, the solid curve is the frequency-dependent renormalization $\Delta'(\omega)$ as obtained from $\Gamma(\omega)$ by a Kramers-Kronig transformation, the straight line (dashed dotted) represents $(\omega - \Omega)$. Ω is the harmonic frequency of the TO mode. (b) Experimental Raman spectrum (----) and calculated spectrum folded with the experimental resolution (——). (KRAUZMANN et al., 1974)

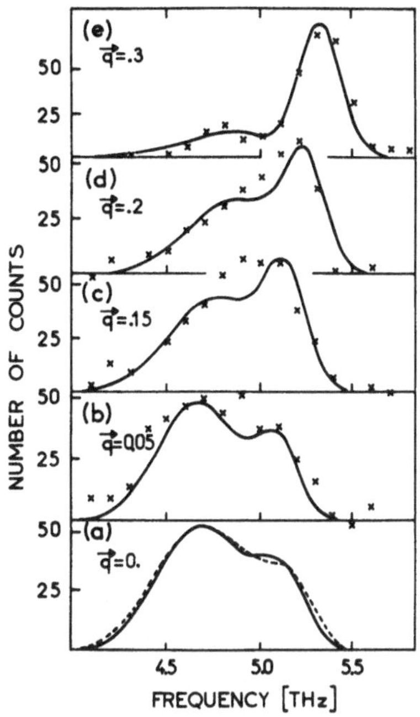

Fig. 47a-e. q dependence of the TO mode in CuCl at 4.2 K in [ξ00] direction. Constant-Q scans (background subtracted) at different wavevectors are compared with model calculations (——). The only q-dependent parameters in these calculations is the harmonic TO mode Ω in (84). The dashed curve at q = 0 is obtained by an average over several experimental scans in different Brillouin zones

We took $\Omega_{TO}(q)$ as measured and kept $\Delta'(\omega)$ and $\Gamma(\omega)$ independent of q and calculated the response for different q values in the [100] direction. The results were convoluted with the resolution of the neutron experiment and compared to experimental data (Fig. 47). Note that $\Omega_{TO}(q)$ is the only varying parameter. The agreement of calculation and experiment is very good despite the poor statistical accuracy of the data.

Regarding Fig. 47 we realize that anharmonicity manifests itself strongly at q = 0 and less with increasing q. For dramatic effects of anharmonicity two things have to come together: i) an anharmonic part in the lattice potential to allow interaction of modes, and ii) a spectrum of decay channels in the area of the single-mode frequency.

7. Final Remarks

Due to our restriction to discuss exclusively lattice dynamics, we have presented the technique of inelastic neutron scattering with correlated restrictions. Basic requirements for the study of lattice dynamics are large energy and momentum transfers. Highly desirable would be very good energy and momentum resolution at the same time.

High resolution in energy would allow separating neighbouring dispersion branches and studying the very interesting field of phonon-phonon interaction (damping of phonons). Currently very little is done in this latter field because at low temperatures, where a theoretical approach is straightforward, the natural linewidths of the phonons are too small to be resolved experimentally, and at elevated temperatures, where the linewidths become measurable, a theoretical description has to include so many "multi" processes that a detailed analysis is not very meaningful. High resolution in momentum would improve the investigation of dispersion curves where the intensity and the frequency varies rapidly with phonon wavevector, as for example at mode anticrossing effects or at critical scattering (at large momentum transfer) near phase transformations.

High resolution always costs intensity. The most economic way to improve resolution is to decrease the energy of the neutrons used. But this collides with the requirement of large energy and momentum transfers. Furthermore sophisticated very high resolution instruments which exist, for example at the Institut Laue-Langevin in Grenoble, work with low-energy (slow) neutrons for technical reasons because 1) neutron guides provide more intensity at large wavelength by the increased angle of total reflection; 2) choppers have a technical limit of highest speed thus giving a limit of shortest pulse length which provides better resolution for slower neutrons; the meaningful distance between two choppers plays a similar role; and 3) the production of polarized neutrons which is necessary for the neutron spin echo technique is more effective at low energies. Such instruments opened up new and very challenging fields of solid-state physics, but they are not compatible with the requirements for the investigation of lattice dynamics.

The flux of continuous neutron sources on the basis of nuclear fission (high flux reactors) cannot be increased drastically due to cooling problems. But pulsed sources using proton accelerators and the effect of spallation from heavy nuclei are expected to provide orders-of-magnitude more flux (in the pulse). The sources will require new technologies in instrumentation and will lead to new fields of research.

A somewhat competing, somewhat compatible project is connected with the very strong X-ray intensity produced by synchrotrons. At synchrotrons it appears feasible to perform inelastic scattering of X-rays with a resolution close to conventional resolution in inelastic neutron scattering. At higher energy transfers (say, above 100 meV) inelastic neutron scattering at continuous sources suffers from the decreasing flux of the Maxwellian spectrum. It is expected that inelastic scattering of X rays at synchrotrons will cover a large range of energy transfer (say, up to 2 eV) with a constant absolute energy resolution (say, 10 meV). It will be interesting to compare the results obtained by inelastic scattering of neutrons, examining the motions of the nuclei, with those from inelastic scattering of X rays which are scattered by the electrons.

References

Abarenkov, I.V.; Heine, V. (1965): Philos. Mag. *12*, 529-537
Aboagye, J.K.; Friauf, R.J. (1975): Phys. Rev. *B11*, 1654-1664
Anderson, P.W. (1960): in *Fisika Dielektrikov*, ed. by G.I. Skanavé (Acad. Nauk SSSR, Moscow) p. 290
Arnold, H. (1965): Z. Kristallogr. *121*, 145-157
Arnold, H. (1976): Habilitationsschrift, Rheinisch-Westfälische Tech. Hochschule, Aachen
Axe, J.D.; Shirane, G. (1970): Phys. Rev. *B1*, 342-348

Barron, T.H.K.; Huang, C.C.; Pasternak, A. (1976): J. Phys. *C9*, 3925-3940
Bauer, K.; Jagodzinski, H.; Dorner, B.; Grimm, H. (1971): Phys. Status Solidi (b) *B48*, 437-443
Bertoni, C.M.; Bisi, O.; Nizzoli, F. (1974): Phys. Lett. *47A*, 466-468
Bilz, H.; Kress, W. (1979): *Phonon Dispersion Relations in Insulators*, Springer Series in Solid-State Sciences, Vol. 10 (Springer, Berlin, Heidelberg, New York)
Birgeneau, R.J.; Kjems, J.K.; Shirane, G.; van Uitert, L.G. (1974): Phys. Rev. *B10*, 2512-2534
Birr, M.; Heidemann, A.; Alefeld, B. (1971): Nucl. Instrum. Methods *95*, 435-439
Bokhenkov, E.L.; Natkaniec, I.; Sheka, E.F. (1976): Phys. Status Solidi (b) *75*, 105-116
Bokhenkov, E.L.; Sheka, E.F.; Dorner, B.; Natkaniec, I. (1977): Solid State Commun. *23*, 89-93
Born, M.; von Karman, T. (1912): Z. Phys. *13*, 297
Born, M.; Huang, K. (1956): *Dynamical Theory of Crystal Lattices* (Clarendon, Oxford)
Boysen, H.; Dorner, B.; Frey, F.; Grimm, H. (1980): J. Phys. *C13*, 6127-6146
Bree, A.; Kydd, R.A. (1970): Spectrochim. Acta *A26*, 1791-1795
Brockhouse, B.N. (1961): In *Inelastic Scattering of Neutrons in Solids and Liquids* (IAEA, Vienna) pp. 113-157
Brockhouse, B.N.; Arase, T.; Caglioti, G.; Sakamoto, M.; Sinclair, R.N.; Woods, A.D.B. (1961): In *Inelastic Scattering of Neutrons in Solids and Liquids* (IAEA, Vienna) p. 531-548
Brovman, E.G.; Kagan, Y.; Kholas, A. (1969): Sov. Phys. Solid State *11*, 733-740
Brovman, E.G.; Kagan, Y. (1974): In *Dynamical Properties of Solids*, ed. by G.K. Horton, A.A. Maradudin (North Holland, Amsterdam) Vol. 1, pp. 191-300

Chaplot, S.L.; Pawley, G.S.; Bokhenkov, E.L., Dorner, B.; Jindal, V.K.; Kalus, J.; Natkaniec, I.; Sheka, E.F. (1981): Chem. Phys. *57*, 407-414
Chesser, N.J.; Axe, J.D. (1974): Phys. Rev. *B9*, 4060-4067
Chernyshov, A.A.; Pushkarev, V.V.; Rumyantsev, A.Yu.; Dorner, B.; Pynn, R. (1979): J. Phys. *F9*, 1983-1995
Chernyshov, A.A.; Pushkarev, V.V.; Rumyantsev, A.Yu., Dorner, B.; Pynn, R. (1981): J. Phys. F (to be published)
Cochran, W. (1959a): Phys. Rev. Lett. *3*, 412-414
Cochran, W. (1959b): Proc. R. Soc. London, *A253*, 260-276
Cochran, W. (1959c): Philos. Mag. *4*, 1082-1086
Cochran, W. (1971): Acta Cryst. *A27*, 556-559
Comes, R.; Shirane, G. (1979): In *Highly Conducting One-Dimensional Solids*, ed. by G.T. Devreese (Plenum, New York) p. 17

Cooper, M.J.; Nathans, R. (1967): Acta Cryst. *23*, 357-367
Cowley, R.A. (1974): Ferroelectrics *6*, 163-178
Currat, R. (1973): Nucl. Instrum. Methods *107*, 21-28

DeWames, R.E.; Wolfram, T.; Lehmann, G.W. (1965): Phys. Rev. *138*, A717-A728
Dick, B.G.; Overhauser, A.W. (1958): Phys. Rev. *112*, 90-103
Di Salvo, F.J. (1977): In *Electron-Phonon Interactions and Phase Transitions*, ed.
    by T. Riste (Plenum, New York) pp. 107-136
Dolling, G.; Waugh, J.L.T. (1965): In *Lattice Dynamics*, ed. by R.F. Wallis (Pergamon,
    New York) pp. 19-32
Dolling, G. (1974): In *Dynamical Properties of Solids*, ed. by G.K. Horton, A.A.
    Maradudin (North Holland, Amsterdam) Vol. 1, p. 541-629
Dorner, B. (1971): J. Appl. Cryst. *4*, 185-190
Dorner, B. (1972): Acta Cryst. *A28*, 319-327
Dorner, B.; Axe, J.D.; Shirane, G. (1972): Phys. Rev. B*6*, 1950-1963
Dorner, B.; Kollmar, A. (1974): J. Appl. Cryst. *7*, 38-41
Dorner, B.; Bauer, K.; Jagodzinski, H.; Grimm, H. (1974): Ferroelectrics *7*, 291
Dorner, B.; von der Osten, W.; Bührer, W. (1976): J. Phys. *C9*, 723-732
Dorner, B. (1976): In *Molecular Spectroscopy of Dense Phases*, ed. by M. Grosmann,
    S.G. Elkomoss, J. Ringeisen (Elsevier Scientific, Amsterdam) pp. 215-224
Dorner, B.; Windscheif, J.; von der Osten, W. (1977): In *Lattice Dynamics*, ed. by
    M. Balkanski (Flammarion Science, Paris) pp. 535-537
Dorner, B.; Comes, R. (1977): In *Dynamics of Solids and Liquids by Neutron Scattering*,
    ed. by S.W. Lovesey, T. Springer, Topics in Current Physics, Vol. 3 (Springer
    Berlin, Heidelberg, New York) pp. 127-196
Dorner, B.; Grimm, H.; Rzany, H. (1980): J. Phys. *C13*, 6607-6612
Dorner, B. (1981a): In *Structural Phase Transformations*, ed. by H. Thomas, K.A.
    Mueller, Topics in Current Physics, Vol. 23 (Springer Berlin, Heidelberg, New
    York) pp. 93-130
Dorner, B.; Bohenkov, E.L.; Chaplot, S.E.; Kalus, J.; Natkaniec, I.; Pawley, G.S.;
    Schmelzer, U.; Sheka, E.F. (1981b): To be published
Dorner, B.; Chernyshov, A.A.; Pushkarev, V.V.; Rumyantsev, A.Y.; Pynn, R. (1981c):
    J. Phys. F*11*, 365-376
Douchin, F.; Lechner, R.E.; Scherm, R. (1981): Nucl. Instrum. Methods, in press
Dutton, D.H.; Brockhouse, B.N.; Miller, A.P. (1972): Can. J. Phys. *50*, 2915-2927

Evans, D.J.; Scully, D.B. (1964): Spectrochim. Acta *20*, 891-900

Fischer, K.; Bilz, H.; Haberkorn, R.; Weber, W. (1972): Phys. Status Solidi (b)
    54, 285-294
Freund, A. (1976): Proc. Conf. on Neutron Scattering, Gatlinburg, p. 1143-1150
Freund, A. (1979): In *Neutron Scattering, Treatise on Materials Science and Techno-
    logy*, ed. by G. Kostorz (Academic, New York) Vol. 15, pp. 461-511
Fujii, Y.; Hoshino, S.; Sakuragi, S.; Kanzaki, H.; Lynn, J.W.; Shirane, G. (1977):
    Phys. Rev. *B15*, 358-368

Gaidukov, Y.N. (1970): Usp. Fis. Nauk *100*, 449-466 [Sov. Phys. Usp. *13*, 194-203
    (1970)]
Garland, G.W.; Silverman, J. (1960): Phys. Rev. *119*, 1218-1222
Gilat, G.; Rizzi, G.; Cubiotti, G. (1969): Phys. Rev. *185*, 971-983
Grimm, H.; Dorner, B. (1975): J. Phys. Chem. Solids *36*, 407-413

Halperin, B.I.; Varma, C.M. (1976): Phys. Rev. *B14*, 4030-4044
Harrison, W.A. (1966): *Pseudo Potentials in the Theory of Metals* (Benjamin, New
    York)
Hennion, B.; Prevot, B.; Krauzmann, M.; Pick, R.M.; Dorner, B. (1979): J. Phys.
    C12, 1609-1624
Hermann, F. (1959): J. Phys. Chem. Solids *8*, 405-418
Horton, G.K.; Maradudin, A.A. (eds.) (1974): *Dynamical properties of Solids* (North
    Holland, Amsterdam)

Ikesawa, M. (1973): J. Phys. Soc. Japan *35*, 309
Ito, M.; Suzuki, M.; Yokoyama, T. (1968): In *Excitons, Magnons and Phonons in Mole-
    cular Crystals*, ed. by A.B. Zahlan (Cambridge, London) pp. 1-29

Jacrot, B. (1970): In *Instrumentation for Neutron Inelastic Scattering Research* (IAEA; Vienna) pp. 225-248
Jeitschko, W. (1972): Acta Cryst. *B28*, 60-76
Johannson, T. (1933): Z. Phys. *82*, 507

Kagan, Y.; Pushkarev, V.V.; Holas, A. (1981): Sov. Phys. JETP in press
Kalus, J.; Dorner, B. (1973): Acta Cryst. *A29*, 526-528
Kalus, J. (1975): J. Appl. Cryst. *8*, 361-364
Kellermann, E.W. (1940): Philos. Trans. *238*, 513
Kitaigorodskii, A.I. (1961): Tetrahedron *14*, 230
Kitaigorodskii, A.I. (1966): J. Chem. Phys. *63*, 9-21
Kleppmann, W.G.; Bilz, H. (1976): Commun. Phys. *1*, 105-110
Kleppmann, W.G. (1976): J. Phys. *C9*, 2285-2293
Kohn, W. (1959): Phys. Rev. Lett. *2*, 393-394
Kovalev, O.V. (1965): *Irreducible Representations of the Space Groups* (Gordon and Breach, New York)
Krainov, E.P. (1964): Opt. Spektrosk. *16*, 984 [English trans.: Opt. Spectrosc. USSR *16*, 532 (1964)]
Krauzmann, M.; Pick, R.M.; Poulet, H.; Hamel, G.; Prevot, B. (1974): Phys. Rev. Lett. *33*, 528-530
Kraxenberger, H. (1980): Thesis, University of Bayreuth, Germany
Kwok, P.C.K. (1967): Solid State Phys. *20*, 213-303

Leigh, R.S.; Szigeti, B.; Tewary, V.K. (1971): Proc. R. Soc. London *A320*, 505-526
Loidl, A.; Daubert, J.; Schedler, E. (1976a): J. Phys. *C9*, L 33-37
Loidl, A.; Daubert, J.; Jex, H.; Schedler, E. (1976b): Phys. Lett. A56, 139-141
Loje, K.F.; Schuele, D.E. (1970): J. Phys. Chem. Solids *31*, 2051-2067
Lovesey, S.W. (1977): In *Dynamics of Solids and Liquids by Neutron Scattering* ed. by S.W. Lovesey, T. Springer, Topics in Current Physics, Vol. 3 (Springer, Berlin, Heidelberg, New York) pp. 1-76
Lovesey, S.W.; Springer, T. (eds.) (1977): *Dynamics of Solids and Liquids by Neutron Scattering*, Topics in Current Physics, Vol. 3 (Springer, Berlin, Heidelberg, New York)
Ludwig, W. (1967): "Recent Developements in Lattice Theory", Springer Tracts in Modern Physics, Vol. 43 (Springer, Berlin, Heidelberg, New York)

Mackenzie, G.A.; Pawley, G.S.; Dietrich, O.W. (1977): J. Phys. *C10*, 3723-3736
Maradudin, A.A., Montroll, E.W.; Weiss, G.H. (1963): "Theory of Lattice Dynamics in the Harmonic Approximation", in Solid State Physics, ed. by F. Seitz, D. Turnbell, H. Ehrenreich (Academic, New York)
Maradudin, A.A.; Vosko, S.H. (1968): Rev. Mod. Phys. *40*, 1-37
Marklund, K.; Vallin, J.; Mahmoud, S.A. (1977): Private communication
Maier-Leibnitz, H. (1967): Ann. Acad. Sci. Fenn. Ser. *A6*, 267
Maier-Leibnitz, H. (1972): In *Neutron Inelastic Scattering* (IAEA, Vienna) p. 681
Michel, K.H.; Naudts, J. (1978): J. Chem. Phys. *68*, 216-228
Migoni, R.; Bilz, H.; Bäuerle, D. (1976): Phys. Rev. Lett. *37*, 1155-1158
Migoni, R.; Bilz, H.; Bäuerle, D. (1978): Ferroelectrics *20*, 157
Minkiewicz, V.J.; Shirane, G. (1970): Nucl. Instrum. Methods *89*, 109-110
Mueller, K.A. (1979): *In Dynamical Critical Phenomena and Related Topics*, ed. by C.P. Enz, Lecture Notes in Physics, Vol. 104 (Springer, Berlin, Heidelberg, New York) p. 210

Natkaniez, I.; Bokhenkov, E.L.; Dorner, B.; Kalus, J.; Mackenzie, G.A., Pawley, G.S.; Schmelzer, U.; Sheka, E.F. (1980): J. Phys. *C13*, 4265-4283
Neto, N.; Scorocco, M.; Califano, S. (1966): Spectrochim. Acta *22*, 1981-1998
Neto, N.; Righini, R.; Califano, S.; Walmsley, S.H. (1978): Chem. Phys. *29*, 167-179
Nicklow, R.M.; Gilat, G.; Smith, H.G.; Raubenheimer, L.J.; Wilkinson, M.K. (1967): Phys. Rev. *164*, 922-928

Ostheller, G.L.; Schmunk, R.E.; Brugger, R.M.; Kearney, R.J. (1968): In *Neutron Inelastic Scattering* (IAEA, Vienna) Vol. I, pp. 315-324

Pawley, G.S. (1967): Phys. Status Solidi *20*, 347-360
Pawley, G.S. (1972): Phys. Status Solidi *B49*, 475-488
Pawley, G.S.; Mika, K. (1974): Phys. Status Solidi *B66*, 679-686

Pawley, G.S.; Mackenzie, G.A.; Dorner, B.; Kalus, J.; Natkaniec, I.; Schmelzer, U.
(1979): Mol. Phys. *39*, 251-260
Phillips, J.C. (1970): Rev. Mod. Phys. *42*, 317-356
Powell, B.N.; Martel, P.; Woods, A.D.B. (1968): Phys. Rev. *171*, 727-736
Powell, B.N.; Dolling, G.; Bonadeo, H. (1978): J. Chem. Phys. *69*, 2428-2434
Prevot, B.; Hennion, B.; Dorner, B. (1977): J. Phys. *C10*, 3999-4011
Price, D.L.; Rowe, J.M.; Nicklow, R.M. (1971): Phys. Rev. *B3*, 1268-1279

Raunio, G.; Rolandson, S. (1970): Phys. Rev. *B2*, 2098-2103
Righini, R.; Califano, S.; Walmsley, S.H. (1980): Chem. Phys. *50*, 113-117
Riste, T. (1970): *Instrumentation for Neutron Inelastic Scattering Research*, (IAEA,
Vienna) pp. 91-98
Riste, T.; Samuelson, E.J.; Otnes, K.; Feder, J. (1971): Solid State Commun. *9*,
1455-1458
Rowe, J.M. ; Rush, J.J.; Chesser, N.H.; Michel, K.H.; Naudts, J. (1978): Phys. Rev.
Lett. *40*, 455-458
Rumyantsev, A.Y.; Pushkarev, V.V.; Zemlianov, M.G.; Parshin, P.P.; Chernoplekov,
N.A. (1978): Zh. Eksp. Teor. Fiz. *75*, 712

Scherm, R.; Dolling, G.; Ritter, R.; Schedler, E.; Teuchert, W.; Wagner, V. (1977):
Nucl. Instrum. Methods *143*, 77-85
Schmelzer, U.; Bohenkov, E.L.; Dorner, B.; Kalus, J.; Mackenzie, G.A.; Natkaniec,
I.; Pawley, S.G.; Sheka, E.F. (1981): J. Phys. *C14*, 1025-1041
Scott, J.F. (1974): Rev. Mod. Phys. *46*, 83-128
Shand, M.L.; Hochheimer, H.D.; Krauzmann, M.; Potts, J.E.; Hansen, R.C.; Walker,
C.T. (1976): Phys. Rev. *B14*, 4637-4646
Sham, L.J.; Ziman, J.M. (1963): In *Solid State Physics* Vol. 15, ed. by F. Seitz,
D. Turnbull (Academic, New York) pp. 221-298
Shapiro, S.M.; Axe, J.D.; Shirane, G.; Riste, T. (1972): Phys. Rev. *B6*, 4332-4341
Shaw, R.W.; Harrison, W.A. (1967): Phys. Rev. *163*, 604-611
Shaw, R.W.; Muhlestein, L.D. (1971): Phys. Rev. *B4*, 969-973
Sheka, E.F.; Bokhenkov, E.L.; Dorner, B.; Kalus, J.; Mackenzie, G.A., Natkaniec, I.,
Pawley, G.S.; Schmelzer, U. (1981): To be published
Springer, T. (1972): "Quasielastic Neutron Scattering for the Investigation of Dif-
fusive Motions in Solids and Liquids", Springer Tracts in Modern Physics, Vol.
64 (Springer, Berlin, Heidelberg, New York)
Stark, R.W.; Falicov, L.M. (1967): Phys. Rev. Lett. *19*, 795-798
Stedman, R.; Almquist, L.; Nilsson, G.; Raunio, G. (1967): Phys. Rev. *163*, 567-574

Taddei, G.; Bonadeo, H.; Marzpochi, M.P.; Califano, S. (1973): J. Chem. Phys. *58*,
966-978
Thoma, K.; Dorner, B.; Duesing, G.; Wegener, W. (1974): Solid State Commun. *15*,
1111-1114
Toepler, J.; Alefeld, B.; Heidemann, A. (1977): J. Phys. *C10*, 635-643
Toigo, F.; Woodruff, T.O. (1970): Phys. Rev. B2, 3958-3966

Vaks, V.G.; Zarochetsev, E.V.; Kravchuk, S.P.; Safronov, V.P. (1978): J. Phys. *F8*,
725-742
Van Dingenden, M.; Hautecler, S. (1963): report BLG 260, CEN-SCK, Mol
Van Vechten, J.A. (1969): Phys. Rev. *182*, 891; *187*, 1007
Vijayaraghavan, P.R.; Niclow, R.M.; Smith, H.G.; Wilkinson, M.K. (1970): Phys. Rev.
B1, 4819-4826
Vissher, P.B.; Falicov, J.M. (1972): Phys. Status Solidi *B54*, 9-42
Von der Osten, W. (1974): Phys. Rev. *B9*, 789-793
Von der Osten, W.; Weber, J. (1974): Solid State Commun. *14*, 1133-1135
Von der Osten, W.; Dorner, B. (1975): Solid State Commun. *16*, 431-434
Vosko, S.H.; Taylor, R.; Keech, G.H. (1965): Can. J. Phys. *43*, 1187-1247

Williams, D.E. (1966): J. Chem. Phys. *45*, 3770-3778
Williams, D.E. (1967): J. Chem. Phys. *47*, 4680-4684
Winston, H.; Halford, R.S. (1949): J. Chem. Phys. *17*, 607-615
Woodruff, T.O.; Ehrenreich, H. (1961): Phys. Rev. *123*, 1553-1559
Woods, A.D.B.; Cochran, W.; Brockhouse, B.N. (1960): Phys. Rev. *119*, 980-999

Zallen, R.; Conwell, E.M. (1979): Solid State Commun. *31*, 557-561

Subject Index

H. Bilz, W. Kress

Phonon Dispersion Relations in Insulators

1979. 162 figures in 271 separate illustrations. VIII, 241 pages
(Springer Series in Solid-State Sciences, Volume 10)
ISBN 3-540-09399-0

Contents:
Summary of Theory of Phonons: Introduction. Phonon Dispersion
Relations and Phonon Models. – Phonon Atlas of Dispersion
Curves and Densities of States: Rare-Gas Crystals. Alkali Halides
(Rock Salt Structure). Metal Oxides (Rock Salt Structure). Transition
Metal Compounds (Rock Salt Structure). Other Cubic Crystals
(Rock Salt Structure). Cesium Chloride Structure Crystals. Diamond
Structure Crystals. – Zinc-Blende Structure Crystals. Wurtzite Struc-
ture Crystals. Fluorite Structure Crystals. Rutile Structure Crystals.
ABO_3 and ABX_3 Crystals. Layered Structure Crystals. Other Low-
Symmetry Crystals. Molecular Crystals. Mixed Crystals. Organic
Crystals. – References. – Subject Index.

G. Leibfried, N. Breuer

Point Defects in Metals I

Introduction to the Theory
1978. 138 figures, 22 tables. XIV, 342 pages
(Springer Tracts in Modern Physics, Volume 81)
ISBN 3-540-08375-8

Contents:
Introduction and Survey. – Harmonic Approximation and Linear
Response (Green's Function) of an Arbitrary System. – Lattice
Theory. – Continuum Theory. – Transition form Lattice to
Continuum Theory. – Scattering of Neutrons and X-rays by
Crystals. – Probability, Distributions and Statistics. – Properties of
Crystals with Defects in Small Concentration. – Appendices. –
References. – Subject Index.

W. Ludwig

Recent Developments in Lattice Theory

1967. 87 figures. VI, 301 pages
(Springer Tracts in Modern Physics, Volume 43)
ISBN 3-540-03982-1

M. Toda

Theory of Nonlinear Lattices

1981. 38 figures. X, 205 pages
(Springer Series in Solid-State Sciences, Volume 20)
ISBN 3-540-10224-8

Contents:
Introduction. – The Lattice with Exponential Interaction. – The
Spectrum and Construction of Solutions. – Periodic Systems. –
Application of the Hamilton-Jacobi Theory. – Appendices A–J. –
Simplified Answers to Main Problems. – References. – Biblio-
graphy. – Subject Index. – List of Authors Cited in Text.

Springer-Verlag
Berlin
Heidelberg
New York